Textbook of

Rainfed Agriculture and Watershed Management

P. L. Maliwal
(Former Director Research, MPUAT, Udaipur)
ICAR-Emeritus Professor

SCIENTIFIC PUBLISHERS

Published by:

SCIENTIFIC PUBLISHERS (INDIA)
5A, New Pali Road, P.O. Box 91
Jodhpur 342 001, India
E-mail: info@scientificpub.com
Website: www.scientificpub.com

Print 2026

ISBN: 978-93-89832-22-8 (HB)
eISBN: 978-93-89832-21-1

Printed in India

डॉ. नरेन्द्र सिंह राठौड़
कुलपति
Dr. Narendra Singh Rathore
Vice-Chancellor

Phone : 91-294-2471101 (O), 2470682 (Fax), 2463839 (R) and 9414166961 (M)
E-mail : vc_mpuat@yahoo.co.in and vc@mpuat.ac.in

Maharana Pratap University of Agriculture & Technology
University Campus, Udaipur-313001 (Rajasthan), INDIA

महाराणा प्रताप कृषि एवं प्रौद्योगिकी विश्वविद्यालय
विश्वविद्यालय परिसर, उदयपुर–313001 (राजस्थान), भारत

No. PS/VC/MPUAT/2020/73
24th January, 2020

FOREWORD

The Indian sub-continent is predominantly characterized by tropical monsoon climate, where rainfall is not only inadequate but unevenly distributed results in frequent droughts and low crop productivity. The new paradigm of dryland agriculture calls for con-current attention to the principles of ecology, economics, equity and employment. In India, about 65 per cent of 143 m ha arable land depends wholly on rainfall contributing nearly 44 per cent food production and support about 40 per cent of human and 60 per cent of livestock population. Productivity of crops in dryland agriculture is low due to inadequate supply of moisture and other constraints. Since production gains from irrigated lands cannot meet the growing demands for ever increasing population of the country more emphasis needs to be given for rainfed agriculture. The population pressure and periodic famine due to unexpected flood and drought in India drawn the attention of scientists to develop suitable technology to get sustainable production from rainfed /dryland agriculture.

A large number of books have been written on dryland agriculture. However, most of them are very elaborative and need is felt to write a book matching with the syllabus of the under graduate students of agriculture. The author presented the book entitled "**Textbook of Rainfed Agriculture and Watershed Management**" in a scientific and systematic manner to understand the basics and fundamentals of the subject clearly and easily. This book is written to fulfill the basic requirement of under graduate students of agriculture stream considering the syllabus recommended by V Dean's committee of ICAR.

I hope that this will serve as textbook for undergraduate students of agriculture stream. I understand that this will also be of equal significance for university teachers, extension personnel dealing in rainfed/dryland agriculture. I congratulate Dr. P.L. Maliwal, Emeritus Professor for his painstaking efforts in bringing out this book including the latest developments in the field of dryland agriculture under the ICAR-Emeritus Professor Scheme.

Looking to its clarity, content, sequence and coverage, I am sure that this book will be widely accepted among the students. I extend my best wishes to Dr. Maliwal for success of this book.

(Narendra Singh Rathore)

Preface

Indian agriculture is characterized by a tropical monsoon climate, where rainfall is not only inadequate but also unevenly distributed results in frequent droughts and crop failures. Moreover, it is considered as gamble of monsoon as production and productivity of *kharif* season crops is mainly governed by quantity and pattern of rainfall. As water is the most crucial single factor in crop production, its scarcity may cause partial or complete failure of crops leading to food scarcity and famines. In India 65 per cent of total arable land (143 M ha) area is rainfed, characterized by low productivity, low income, low employment and marginal lands. Rainfall pattern in these areas are highly variable both in terms of total amount and its distribution. Taking the country as a whole, as per meteorological report, severe drought in large areas is experienced once in 50 years and partial drought once in five years while floods are expected every year in one part of the country or the other. This drawnd the attention of the scientists in the country to develop technology befitting for dryland/rainfed agriculture to get sustainable production.

As entire process of agricultural development depends on the status of water resources, watershed with distinct hydrological boundaries is considered ideal for planning developmental programmes. It is essential to have various developmental programmes on watershed basis in conjunction with basic soil and water conservation measures.

A large number of books are available on rainfed agriculture and watershed management but most of them are much elaborated and it is difficult for the students to find out the relevant materials as per their syllabus. Therefore, efforts have been made to compile and present the information available on rainfed agriculture and watershed management in a cogent and easily understanble manner. The book is written basically to fulfill the requirement of undergraduate students of agriculture stream considering the syllabus of Vth Dean's Committee of ICAR. The Author is optimistic that the students will get the benefit of this book in understating the fundamentals of rainfed agriculture and watershed management.

The author is deeply indebted to Indian Council of Agricultural Research, New Delhi for providing me the opportunity for preparing the manuscript under the Emeritus Professor Scheme of ICAR.

The Author acknowledges his indebtness to authors of the books and other publications from which most of the material/literature have been sought. In several cases, it has not been possible to obtain permission for reproduction, for which the author/publisher offer apologies.

The author is thankful to Dr A. Joshi, Dean, RCA, Udaipur and Dr. Dilip Singh, Professor and Head Deptt. of Agronomy for their generous support in bringing out this publication.

The author is also thankful to Dr. P.K. Singh, Professor, Soil and Water Conservation, CTAE, MPUAT, Udaipur and Dr. S. K. Sharma, Zonal Director Research, MPUAT, Udaipur and faculty members of Department of Agronomy, RCA, Udaipur for the help and suggestions rendered for betterment of this book.

The author welcomes the constructive comments and suggestions from the readers whosoever for further improvement of this book and that will be duly acknowledged. I sincerely thank to Scientific Publishers, Jodhpur for publishing this book well in time.

P.L. Maliwal

Contents

1

AN INTRODUCTION TO RAINFED AGRICULTURE

Agriculture is the single largest livelihood source in India with nearly two third of the people depend on it. Indian agriculture is predominately a rainfed agriculture under which both dry farming and dryland agriculture is included. Most of the cropping in arid and semi-arid regions continues to be under rainfed conditions. In India most of the dryland/rainfed areas are resource constrained. Rainfed agriculture is practiced under a wide range of soil type, agro-climate and rainfall conditions. Crops raised in these conditions are prone to the monsoon failures, variable in rainfall quantity and its distribution, diversity in crop management practices and variability of the soil type. The prolonged moisture stress may lead to partial or complete failure of crops. With the limited scope to increase area under plough only option left is to enhance the productivity of rainfed areas. The significance of dryland/rainfed farming in India is described as under.

1.1 Significance of Dryland/Rainfed Farming

Dryland agriculture is one, which is extreme to irrigated agriculture in terms of moisture availability with more than 60 per cent risk in crop production.

In India both dryland and rainfed agriculture are being practiced with 30 to 40 per cent productivity as compared to irrigated agriculture. Rainfed agriculture is similar to irrigated agriculture in all activities except the rainfall which is the main source of moisture supply; however, water logging may be the problem in many circumstances. In India about 65 per cent is dryland / rainfed agriculture and contribute significantly in food grain as well as fodder production. Significance of dryland farming in Indian agriculture is due to following facts:

(i) Around 70 per cent of rural population lives in dryland areas and their livelihood depends on success or failure of the crops.

(ii) Dryland agriculture plays a distinct role in Indian agriculture occupying about 100 Mha cultivated area, supports 40 per cent of human population and 60 per cent of livestock population.

(iii) Dryland farming comprises about 91 per cent area of coarse cereals (sorghum, pearlmillet, maize and fingermillet), 91 per cent of pulses (chickpea, pigeonpea, etc.), 80 per cent of oilseeds (groundnut, rapeseed & mustard and soybean) and 65 per cent of cotton. Also about 50 per cent area under rice and 19 per under wheat is rainfed.

(iv) At present three hactare of dryland crop produce cereals grain equivalent to that in one hectare irrigated area. With limited scope for bringing more area under cultivation, only option left is to increase the productivity of crops using improved technology.

(v) The productivity of food grains also reached to a platue in irrigated areas due to problems of nutrient exhaustion, salinity buildup and increase in water table in canal irrigated areas. Therefore, the challenge of future would be to produce more from drylands using improved technology and efficient use of resources.

(vi) It is anticipated that evolving new technology for dryland agriculture will make fragile dryland ecosystem more productive as well as sustainable. In order to achieve evergreen revolution, we shall have to make grey areas (dryland) green through new technological innovations.

(vii) Dryland offer good potential for adoption of alternate land use systems like agroforestry, alley cropping, ley farming, etc. which will not only supply food, fuel to the village people and fodder to the cattle but also conserve the natural resources for ecological balance.

1.2 Types of Dryland Agriculture

Usually, the terminology dry farming, dryland farming and rainfed farming are used synonymously to indicate similar farming situations. However, depending on the amount of rainfall received, dryland agriculture can be categorized in three groups as follows:

(i) **Dry Farming** – Cultivation of crops in areas where rainfall is less than 750 mm per annum is called dry farming. Prolonged dry spells during crop season are most common. Crop failures are more frequent. Moisture conservation practices are common in these areas (arid regions).

(ii) **Dryland Farming** – Cultivation of crops in areas receiving rainfall between 750 and 1150 mm per annum is known as dryland farming. Dry spells during crop season occurs, but crop failures are less frequent. In semi-arid regions moisture conservation practices and provision of drainage especially in black soils are necessary.

(iii) **Rainfed Farming** – Cultivation of crops in regions receiving rainfall more than 1150 mm per annum. It is practiced in sub-humid and humid regions where crop failures are rare and drainage is the major problem.

United Nations Economic and Social Commission for Asia and the Pacific classified dryland agriculture into two categories, dryland farming and rainfed farming on the basis of the criteria given in table 1.1 as under.

Table 1.1: Dryland Farming v/s Rainfed Farming.

Constituent	Dryland farming	Rainfed farming
Rainfall (mm)	45.5% (65.5 mha) >750 mm/annum	24% (34.5 mha) >1150 mm
Moisture	Shortage	Enough
Growing region	Arid and semi-arid and uplands of humid and sub-humid regions	Humid and sub-humid regions
Cropping pattern	Single crop or intercropping	Intercropping or double cropping
Growing season (days)	< 200	> 200
Constraints	Scarcity of water, wind and water erosion	Water erosion and drainage

If we look the above comparison between dryland farming and rainfed farming, it is clear that in both the situations there is no provision for irrigation or both exclude irrigation. Further it also becomes evident that the emphasis in rainfed farming is often on disposal of excess water, maximum crop yields, high level of fertilizer inputs and water erosion constraints. While in dryland farming, emphasis is on water conservation, sustainable crop yield, limited inputs for soil fertility maintenance and both wind and water erosion constraints.

1.3 Characteristics of Dry Land/Rainfed Agriculture

Rainfed agriculture in India is under a wide range of climate and soil conditions, the main characteristics features are given as under:

(i) **Uncertain, Ill Distributed and Limited Annual Rainfall:** Based on average annual rainfall the country can be divided into three zones; low (<750 mm), medium (750–1150 mm) and high (>1150 mm) rainfall zones. The rainfed area is nearly equally distributed in the above three zones in India. Nearly 75 per cent of annual rainfall is received during rainy season, *i.e.* June to September. For successful crop production, distribution of rainfall is more important than the quantum, which is uncertain and ill distributed in the country. This leads to drought conditions in some areas and water logging in another areas, ultimately, adversely effects the productivity in both the situations.

(ii) **Undulating Soil Surface:** As rainfall is unevenly distributed, there is soil erosion by both water and wind. During rainy season there is water erosion in some areas due to excess rainfall, while in other areas there is scarce rainfall leads to wind erosion. Therefore, every year soil is eroded in one or other way leads to formation of gullies or rills, require levelling of land. However, due to poor economic conditions of farmers, required land levelling is not done, ultimately leads to low productive undulating land.

(iii) **Extensive and Large Farm Holdings:** In drought prone areas usually the density of human population is less and farmers have large holdings. As the farmers are resource poor and hesitate to adopt improved technology due to uncertainty of rains, leads to poor average productivity of crops.

(iv) **Practice of Extensive Agriculture:** The type of farming adopted in dryland areas is usually extensive due to lack of irrigation facilities. In these areas usually monocropping or intercropping is practiced during *kharif* / rainy season and in *rabi* there is no crop. Some time farmers keep the field fallow during *kharif* season, conserve the moisture and take *rabi* crop. Thus the cropping intensity remains only 100 per cent in these areas.

(v) **Monotonous Cropping in a Particular Region:** In dry farming areas farmers hesitate to change the cropping pattern due to low risk bearing capacity of dryland farmers. They use to grow drought resistant crops like pearlmillet, sorghum, maize, minor millets, pigeonpea, kidneybean, greengram, blackgram, sesamum, etc during, *kharif* season and chickpea, lentil, mustard, toria, taramira, linseed, etc. during *rabi* season depending on the rainfall pattern of that particular area. They hesitate to introduce any other high value crop. Thus the entire region can be seen under same crop/cropping pattern having poor growth and development, and can be identified as rainfed/dryland areas.

(vi) **Low Crop Productivity:** The crop yields in dry land areas is quite low as farmers being recourse poor and do not follow the recommended package of practices, such as adoption of improved crop varieties, balanced fertilizer use, plant protection measures, weed control, etc. Farmers do not want to apply any monetary inputs, as it is very risky and many times it does not meet the expenses incurred on sowing and harvesting of crops due to failure of rains and prolonged drought situations.

(vii) **Poor Market Facility for Sale of the Produce:** As most of the farmers in a particular region use to grow same crops and they become ready at the same time, hence the farmers have to face a hard time and they get low price of their produce because of poor demand and abundant supply rather a glut in the market. Very often the people of the locality do not want to buy the commodity, and farmers are forced to sell their produce to local traders, in want of money and fetch very low prices. At this times Govt. is also not in position to purchase the large produce at minimum support price (MSP) and ultimately farmers do not make any profit.

(viii) **Poor Economy of Farmers:** Dryland farmers are usually resource poor and their economic conditions is not good. They do agriculture for securing their livelihood as they do not have any alternate employment opportunity. Due to scarcity of fodder and feed the cattle are not stale fed, they mostly depend on grazing, ultimately low productivity. The poor economy of the farmers affects their livings as well as action and brought up of their children.

1.4 Concept of Dryland Agriculture

In earlier concept the areas receiving annual rainfall of less than 500 mm were considered as dry farming. However, the concept of dryland/rainfed agriculture (farming) has been changed now. In modern concept, dryland areas are those where the balance of moisture is always on the deficit side. In other words, annual potential evapo-transpiration (PET) exceeds precipitation in dryland/rainfed agriculture; there is no consideration of amount of rainfall. It may appear quite strange to a layman that even those areas which receive 1100 mm or more rainfall annually fall in the category of dryland agriculture under this concept. To be more specific, the average annual rainfall of Varanasi is around 1100 mm and the annual potential evapo-transpiration is 1500 mm, thus the average moisture deficit so created comes to 400 mm. This deficit in moisture is bound to affect the crop production under dryland situations, ultimately resulting into total or partial failure of crops. Accordingly, production is either low or extremely uncertain and unstable which are the real problems of dryland in India.

Success of crop production in these areas depends on the amount and distribution of rainfall, as these influences the stored soil moisture and moisture used by crops. Amount of water used by the crops and stored in soil is governed by water balance equation: $ET = P - (R + S)$.

Where ET is evapo-transpiration, P is precipitation, R is runoff and S is soil moisture storage. When balance of the equation shifts towards right, precipitation (P) is higher than ET, so that there may be water logging or it may even lead to runoff (R) and flooding. On the other hand, if the balance shifts to left, ET becomes higher than precipitation, resulting in drought.

Taking the country as a whole, as per meteorological report, severe drought in large areas is experienced once in 50 yr and partial drought once in five years, while floods are expected every year in one part of the country or the other, especially during rainy season. In fact the balance of the equation is controlled by weather, season, crops and cropping pattern.

1.5 History of Dryland /Rainfed Agriculture in India

Dry farming is a practice of antedates history, but use of the term in its present form and meaning probably started in Utah in 1863. Dr John A. widtsol one of the earliest to take up dry farming research defines the term as the profitable production of useful crops without irrigation on lands that receive annually a rainfall of 500 mm or less, this he says may be extended to 750 mm with unfavourable distribution of rainfall. According to definition given in the Encyclopedia Britannica "Dry farming is farming without irrigation in areas where the moisture supply is the chief limitation to crop production". Dry farming and dryland farming have been used synonymously to indicate similar farming situations.

In India agriculture is considered as gamble of monsoon as the production and productivity of *kharif* season crops is mainly governed by quantity and pattern of rainfall. From time immemorial the chief form of agriculture in dryland tracts of India was cultivation of drought tolerant millets for food and fodder. As water is the most important single factor in crop production, its scarcity may cause partial or complete failure of crops leading to periodic food scarcity and famines. To address these issues, the government of India appointed the first Finance Commission in 1880. The Commission recommended creation of protective irrigation projects in dryland areas of the county.

First systematic scientific approach to tackle the problems of dry farming areas was initiated by Tamhane in 1923 on a small plot at Manjri Farm near Pune and the work passed on to Kanitakar in 1926. A comprehensive scheme of research was drawn up by Kanitakar with financial support from the Indian Council of Agricultural Research (ICAR). Realizing the importance, ICAR launched a comprehensive project on dryland farming at five centres; Sholapur and Bijapur in 1933, Hagari and Raichur in 1934 and Rohtak in 1935. A decade of work upto 1933-44 mainly on rainfall analysis, Physico-chemical properties of soils, physiological studies on millets and on agronomic aspects resulted in a series of improved farming practices for dryland areas commonly known as the **Bombay dry farming** practices. These practices stressed the need for contour bunding, deep ploughing, application of farmyard manure, low seed rate with wider spacing, mixed cropping and crop rotation. These recommendations could not motivate the farmers to adopt them as the yield advantage was only around 15-20 per cent over the base yield of different crops.

By the mid 1950s, the emphasis was shifted to soil management. Soil Conservation Research and Training Centres were established in eight locations in India. The programme's focus was on contour bunding. However, such an effort sometimes gave negatives results, particularly in vertisols. Water congestion and washout were the main problems. Even when yield increases were observed, they were again not more than 15-20 per cent over the base yield.

The importance of shorter duration crops to fit into the moisture availability period was recognized during 1960s and it was during the mid -1960s that the answer was available through hybrids and high yielding varieties of crops, particularly of coarse cereals. These high yielding hybrids and varieties were not only responsive to fertilizers but also to management practices.

Through the establishment of All India Coordinated Research Project on Dryland Agriculture (AICRPDA) in 1970, the emphasis was shifted to multi-disciplinary approach to tackle the problems from several angles. Similar efforts were initiated at the International Crop Research Institute for Semi Arid Tropics (ICRISAT) at Hyderabad during 1972. A brief history about the research efforts on dryland farming in India is given as under in table 1.2.

Table 1.2: History of Dryland Research in India.

Year	Research Institutes/Projects
1923	Dryland Research Station at Manjri, Pune
1933	Research Stations established at Bijapur and Solapur
1934	Research Stations at Hagari and Raichur
1935	Research Station established at Rohtak (Punjab)
1950	Soil Conservation Research and Training Centres established at eight locations in India
1970	23 Research Centres established under All India Coordinated Research Project on Dryland Agriculture (AICRPDA), Hyderabad
1972	Establishment of International Crop Research Institute for Semi Arid Tropics, (ICRISAT), Hyderabad
1976	Establishment of Operational Research Project (ORP) on Dryland Agriculture, Hyderabad
1983	Start of 47 Model Watersheds under ICAR
1985	Birth of Central Research Institute for Dryland Agriculture (CRIDA), Hyderabad
1988	National Research Centre for Agroforestry, Jhansi
1990	National Research Centre for Arid Horticulture, Bikaner
2006	National Rainfed Area Authority (NRAA), New Delhi

1.6 History of Watershed Development in India

In India about 65 per cent of total arable land (143 Mha) area is rainfed, characterized by low productivity, low income, low employment and marginal land. Rainfall pattern in these areas is highly variable both in terms of total amount and its distribution, which leads to moisture stress during critical stages of crop production and makes agriculture production vulnerable. To manage these areas watershed approach was considered most appropriate to make best use of available land and water resources.

Watershed Development Projects (WDP) in the country has been sponsored and implemented by Government of India from early 1970s. Various watershed development programmes were launched subsequently in various hydro-ecological regions. Entire watershed development programme was primarily focused on structural - driven compartmental approach of soil conservation and rain water harvesting during 1980s and prior to that.

The Intergrated Watershed Development Programme (IWDP) with participatory approach was emphasized during mid 1980s and in early 1990s. This approach had focused on raising crop productivity and livelihood improvement in watershed along with soil and water conservation measures. The Government of India appointed a committee in 1994 under the chairmanship of Prof C.H. Hanumantha Rao. The committee thoroughly reviewed existing strategies of watershed programme and strongly felt a need for moving away from the conventional approach of the government department to the bureaucratic planning without involving local communities. The new guidelines were recommended in 1995, which emphasized on collective action and community participation, including participation of primary stakeholders through community based organizations, non-governmental organizations and Panchayati Raj Institution (PRI).

Need for unification of multiplicity of watershed development programmes within the frame work of a single national initiative was felt in the year 2001. A sub-committee constituted for this purpose suggested common guidelines for watershed development projects through National Authority for Development of Rainfed Areas (NADRA). This finally resulted guidelines for watershed development (Neeranchal) were released in 2008. These common guidelines were revised in 2013. Chronology of Watershed Development (WSD) programmes in India are given below in table 1.3:

Table1.3: Watershed Development Programmes in India

Year	Programme
1973-74	Droughts Prone Area Programme (DPAP)
1977-78	Desert Development Programme (DDP)
1989-90	Intergrated Wasteland Development Programme (IWDP)
1989	Intergrated Afforestation and Eco-development Scheme (IAEPS)
1990-91	National Watershed Development Project for Rainfed Areas (NWDPRA)
2001	Common Guidelines for Watershed Development (revised)
2002	National Afforestation Programme
2003	Hariyali Guidelines
2008	Common Guidelines for Watershed Development (Neeranchal)
2009	Intergrated Watershed Management Programme (IWMP)
2011	Revised Common Guidelines for Watershed Development
2013	Revised Common Guidelines (known as Neeranchal Guidelines)

1.7 Projects/Institutes Running for Development of Dryland Agriculture

1.7.1 All India Coordinated Research Project on Dryland Agriculture (AICRPDA)

AICRP on Dryland Agriculture was established by ICAR during the year 1970 at Hyderabad with the assistance of Canadian International Development Agency (CIDA), has the net work of 23 centres and 8 Operational Research Projects (ORPs) spread across arid, semi-arid, sub-humid and humid rainfed agro-ecologies. The primary goal of this project is to stabilize and increase agricultural production in drylands of India. The AICRPDA is a multi-disciplinary project encompassing the disciplines of Agronomy, Soil Science, Plant Breeding and Agricultural Engineering.

The cooperating cell of the project was located at Hyderbad, Andhra Pradesh. To begin with, there were 15 main centres and 8 sub-centres. In 1984, 5 out of 8 sub-centers were upgraded to the level of main centre. An additional special centre (twenty fourth) was set up at Indian Agricultural Research Institute, New Delhi, with the hope that it will carry out path breaking research in the fields of Plant Physiology and Agro - meteorology. Lately it was merged with coordinating cell of the project. In 1985 Dehradun centre was dropped.

1.7.2 Operational Research Project (ORP)

The first Operational Research Programme on Dryland Agriculture was launched at Indore (Madhya Pradesh) with U.K. collaboration. It was augmented with the addition of four more CIDA (Canadian International Development Agency) sponsored ORPs in 1976. In 1984, four additional ORPs were added to the list. In all nine dryland Operational Research Projects were operating at Indore, Bangalore, Hyderabad, Sholapur, Ranchi, Hoshiarpur, Anantpur, Hisar and Arjia, Bhilwara till 2017-18. From 2018-19 the project was merged with AICRPDLA giving additional posts of one scientist and two supporting staff to initiate work on rainfed integrated farming system on farmers field.

1.7.3 Central Research Institute for Dryland Agriculture (CRIDA)

In recognition of commendable research work done by the scientists of AICRPDA and in view of the national urgency and importance of developing more stable and remunerative farming systems for drylands of the country, the Indian Council of Agricultural Research elevated the status of the Directorate of AICRPDA to that of all full fledged institute, as CRIDA, Hyderabad. On April 19, 1985 the institute was formally inaugurated. The main purpose of establishing CRIDA, was to carry out lead research in the area of dryland agriculture, leaving location specific problems and their solutions to AICRPDA. The Central Research Institute for Dryland Agriculture presently operates three projects -

(i) All India Coordinated Research Project on Agro-meteorology (AICRPAM).

(ii) All India Coordinated Soil Test Crop Response (STCR) Correlated Project.

(iii) All India Coordinated Project on Dryland Agriculture (AICRPDA)

1.7.4 International Crop Research Institute for Semi Arid Tropics (ICRISAT)

The International Crop Research Institute for Semi Arid Tropics was established at Hyderbad, Andhra Pradesh. A meaningful collaboration between CRIDA and ICRISAT on one to one scientist basis started from 1986 onwards in areas of biological nitrogen fixation, agroforestry systems, land configurations for soil & moisture conservation and economics of watershed technology. CRIDA and ICRISAT also joined hands in management of dryland agriculture on watershed basis. Additional collaboration was initiated in 1989 on evaluating the vegetative barriers, studying the behavior of phosphorus on vertisols and investigation on perennial pigeonpea. Currently ICRISAT is engaged in research in India on a major scale in pearlmillet, chickpea, groundnut, pigeonpea and sorghum. The results of ICRISAT are gainfully been utilized by CRIDA also in development of dryland agriculture in India.

1.8 Problems of Dryland Agriculture

The major problems associated with dryland/rainfed agriculture are as follows:

(i) **Moisture Stress due to Limited and Uneven Distribution of Rainfall:** The rainfall received in dryland areas is not only scarce but very erratic/ uncertain and poorly distributed, ultimately leads to drought/waterlogging conditions results in poor crop yields. Therefore, in India agriculture becomes a sort of gamble of monsoon and in these situations success of crop is not certain and farmers get very nominal crop yields or sometimes no yield. Farmers have to face aberrant weather conditions like late onset of monsoon, mid season drought and early withdrawal of monsoon.

(ii) **Lack of Effective Storage of Rain Water:** The most important feature of dry land farming is either there will be scarce rainfall or there will be torrential rain with very high intensity. Thus in the former case the crops have to face severe drought and in latter case they will face water logging, ultimately the yield is severely affected. In our country farmers neither properly adopt *in-situ* moisture conservation technology nor they have water storage facility for excess rain water, to reuse it as life saving irrigation.

(iii) **Lack of Improved Technology:** In India there is lot of variability exists in crops and cropping pattern, rainfall pattern, soil type and other climatological conditions. However, very limited research is done on different aspects like development of suitable varieties of different crops for different situations, soil and water conservation technology feasible for the farmers, weather forecast, nutrient management, water harvesting and recycling, alternate land use, etc. Very limited research has been carried out in All India Coordinated Research Projects on Dryland Agriculture, CRIDA, Hyderabad.

(iv) **Limited Crop Choice for Dryland Areas:** In dryland farming only drought tolerant crops like coarse cereals & millets, pulses and oilseeds can be grown. The choice of the farmers is very limited and cannot afford to grow high value crops. Moreover, they have to purchase rest of the commodities from the open market to fulfill their household needs.

(v) **Limited Use of Fertilizer and Organic Manures:** Farmers use to apply organic manures mostly in irrigated crops only. Farmers also hesitate to use as basal dressing of N, P and K fertilizers due to uncertainty in success of crops. Mostly they apply nitrogenous fertilizers as top dressing. This ultimately leads to low crop productivity even if rainfall is normal. Use of bio-fertilizers is also very rare.

(vi) **Non Availability of Improved Variety Seed:** Short duration drought tolerant varieties seed is not available as and when required by the farmers to cope up with aberrant weather conditions. For different rainfall patterns different crops and their varieties are recommended for dryland conditions. However, when such condition prevails the seed of desired crops variety is not available in the market as seed bank concept does not prevailed in the country. Ultimately farmers are forced to grow the crop and varieties as per availability in the market, which gives sub-optimal yields.

(vii) **Lack of Precise Weather Forecast:** In dryland agriculture weather forecast is very much deciding factor in failure or success of crops as it can guide the farmers in planning the cropping pattern and also plant protection strategy. However, the accuracy in medium and short range forecast at grass root level is still not very high to face the vagaries of monsoon in drylands.

(viii) **Inadequate Extension Activities:** Popularization of dryland technology among the farmers is also inadequate. Farmers are not aware of new innovations in dryland farming to stabilize the production.

(ix) **Poor Economic Conditions:** Most of the dryland farmers are resource poor which make them avoiding risk. They do not hire labour even for critical farm operations and engage only family labour force for most of the farm activities. Due to poor financial status of farmers they avoid to use costly seeds, fertilizers, plant protection measures, etc. as their risk bearing capacity is very low.

(x) **Financial Problem:** Lack of capital, support price risk for the produce, marketing facilities and credit facilities at proper time make the farmers hesitant to adopt the improved dryland technologies.

(xi) **Lack of Farm Mechanization:** Dryland farming is practiced in *kharif* season where there is uncertainties of monsoon and farmers get a narrow window for different farm operations from sowing to harvesting. However, due to unavailability of efficient farm machines at village level, the peak operations are not perfomed in time, have very negative impact on production of crops.

1.9 Future Prospects

The productivity gains in crops has already showed a plateau in irrigated agriculture due to problems related to nutrient exhaustion, salinity build up and rising water table. Therefore, the challenges of the present millennium would be to produce more from drylands while ensuring conservation of existing resources. Hence, new strategies would have to be evolved which would make the fragile dryland ecosystems more productive as well as sustainable. In order to achieve evergreen revolution, we have to make grey areas (drylands) green through adoption of improved technology.

In India only 43 Mha of land out of the cultivated area of 143 Mha was irrigated leaving the remaining 100 Mha as rainfed. According to experts, even when the ultimate irrigation potential is reached, 55 per cent of the total cultivated area will be still, rainfed. Therefore, for sustainable production from dryland agriculture, tremendous efforts both in the development and research in rainfed /dryland agriculture are essential on following important aspects:

(i) ***In-situ* Moisture Conservation:** There is need to develop improved methods/technologies for *in-situ* moisture conservation as crops are wholly dependent on rain water. Ongoing bunding programmers for moisture conservation, vegetative barriers on contours can not only be economically viable alternatives to earthen bunds for soil conservation, but can also act as effective key lines for contour cultivation. Identification of locally available grasses and suitable methods of their establishment needs to be studied on priority basis.

(ii) **Nutrient Management:** No doubt the consumption of fertilizers in the country has drastically increased, but about 80 per cent of it is used in crops grown under irrigated conditions. However, there is a good response to the application of fertilizers under dryland conditions which is as high as 23 kg per ha of nutrient under favourable situations. One of the sure ways to increase the yield under dryland conditions is to step up the consumption of fertilizers at least to about 50 kg total nutrients per ha.

(iii) **Soil Organic Matter:** Increasing and maintaining soil organic matter becomes almost impossible especially under tropical conditions when the crop residue is needed both for feeding the animals and for fuel purposes. Under extreme conditions, the soil organic matter content drops to the point where the soil become useless and soil erosion becomes rampant. The alternative is to produce biomass from the waste lands in the arid and semi-arid conditions. Identifying suitable vegetation and the methods of their establishment should be a priority item in the future of our dryland research. Covering the wasteland with vegetation will be inevitable to reduce the soil erosion.

(iv) **Water Harvesting and Recycling:** Water harvesting is one of the most promising and technically feasible recommendations even in low rainfall areas where water is lost as runoff. At the same time the methods now suggested for seepage control, lifting of water for irrigation are often uneconomical, therefore, there is need to divert research in the directions to evolve/select cost effective sealants and alternation of manually operated lifting devices.

(v) **Development of Farm Implements:** There is urgent need to develop cost effective implements/labour saving devices for land preparation, sowing and intercultivation which can improve the crop yields by achieving the much talked timeliness and precision in operations under dryland situations. Implements operated by human labour or single bullock are likely to be more acceptable to the resource poor dryland farmer.

2

SOIL AND CLIMATE

Rainfed farming in India is practiced under wide range of climatic and soil conditions. The soils of drought prone area vary from very shallow to, medium deep and deep. On account of more or less complete absence of leaching the soils are base saturated. The exchangeable calcium is predominant cation and the free lime reserve is fairly high (3 to 10%) in these soils. The problematic soils viz. saline, saline sodic do occur in patches in low lying areas. As regard the fertility status, the soils are generally low in organic carbon (0.35 to 0.5%), total nitrogen (0.03 to 0.05%), low to medium in available phosphorus (10 to 30 kg /ha) and high in available potash (300 to 750 kg/ha). Usually micro nutrient deficiencies are not observed in dryland crops.

Further soils exhibit varying degree of erosion depending on the soil slope, tillage operations and cropping season. The soils classed as moderate to high erodible. Hence soil and water conservation is a pre-requisite for successful cropping. Limited soil depth puts limitations on availability of water and nutrients for the crops grown. Usually soils having less than 45 cm depth are suitable for *kharif* crops only as they are unretentive of soil moisture. Inter-mittant wetting due to frequent rainfall during June to August helps to mature crops on such soils. Soils having depth more than 45 cm have high moisture storage and retentive capacity, suitable for *rabi* sowing or double cropping.

2.1 Characterization of Soils of Drylands

The characteristic features of the major soil groups and prospects of raising rainfed crops are described as under:

2.1.1 Vertisols

These soils are commonly called **black soils,** occupying around 20 per cent of the geographical area in the country. It carries the greatest potential of crop production under rainfed conditions. Rainfall in these areas varies from 500-1500 mm. These soils are characterized by high clay content (30-70%) and high water holding capacity. These soils have large variations in soil depth (15-240 cm), low in nitrogen and phosphorus contents. These soils have physical constraints such as narrow range of soil water content for tillage, tendency to become waterlogged and poor traffic ability. As soil organic matter is low in Indian vertisols, the productivity is largely dependent on clay content. Land degradation from soil erosion and salt accumulation, especially in low lying areas is also a constraint in crop production.

2.1.2 Alfisols

These soils are commonly referred as **red soils,** occupying about 20 per cent of geographical area. Annual rainfall varies between 750–2000 mm. These soils are characterized by moderate clay content and poor in nutrient status especially nitrogen, phosphorus, sulphur and calcium. They have low water holding capacity due to shallow depths. The soil have (compact) sub - soil strata prone to large volume of runoff leads to soil erosion. Problem relating to crop production in these soils are poor crop stand due to crusting and rapid drying of surface soil and poor crop growth due to unreliable soil moisture supply.

2.1.3 Aridisols

These soils occupying nearly 9 per cent of geographical area in the country. The rainfall in these areas is also very low. These soils are extremely light *i.e.* sandy, loamy sand or sandy loams, having very low clay content. Calcium carbonate occurs at various depths influencing the effective depth of the soil. Crusting tendency of these soils is a serious problem. The thickness of the solum varies from 70–100 cm. As the soils are sandy it has low water retention capacity and poor soil fertility. They are universally poor in nitrogen and sometimes inadequate even in phosphorus. Sub soil salinity is not common due to extreme aridity in deep sandy loams either a rainy or post rainy crop can be raised. But in loamy sands or sandy loams, only rainy season cropping is possible. Crop failures are common. Better suited to alternate land use.

2.1.4 Inceptisols and Entisols

These are commonly termed as **alluvial soils,** occupies about 21 per cent of geographical area in the country. These soils are very deep having variable texture and generally deficient in nitrogen and phosphorus contents. The soils possess high production potential. They have low water holding capacity. Management of these soils for crop production is relatively easy compared to red and black soils. Soil erosion is however, a problem leading to land degradation.

2.1.5 Sub-montane Soils

These soils are spread in hill and foot hill regions of the country. These soils experience an annual rainfall varying between 750–2000 mm and highly prone to erosion and runoff losses, on account of topographic features. These soils are medium in texture and depth is medium to deep as well as moderate in clay content. Moisture holding capacity is high (300 mm m^{-1} profile depth). Due to high rainfall double cropping is possible in these soils.

2.2 Climate Characterization of Drylands

Weather, which is part of climate, is one of the major factors influencing the growth, sustenance and crop yields. In fact, the influence of weather is operative even before the crop is sown; the effect of weather on the crop in its total life cycle has been well recognized. In spite of the cultivation of high yielding varieties, improved cultural practices and plant protection measures, favourable weather is a must for good harvests. On the other hand, weather abnormalities such as cyclones, droughts, hailstorms, frost, high winds, extreme temperature and insufficient photosynthetic radiation, etc. may sometimes lead to very low or even no yield at all. Hence, characterization of agro-climates is a pre-requisite for analyzing the agricultural potential of a region, more so under dryland agriculture. Agroclimatic characterization of the dryland ecosystem holds great importance in agriculture planning and management of Indian agriculture. Such a characterization helps in developing strategies for improving and stabilizing the dryland productivity.

India is a country with varied climates resulting mainly due to the spatial and temporal variability in rainfall and temperature conditions. Though the dryland agriculture is practiced from deep south to extreme north, the climate contrasts are striking. Southern Tamil Nadu experience typical tropical temperatures with north east (NE) monsoon being the main source of rainfall. In contrast, Punjab and Haryana in the north-west experience continental climates with extremes of temperatures varying from 45-50°C in summer and near freezing temperatures in winter and dryland cultivation in these areas is restricted to the south west (SW) monsoon season. Based on the rainfall and temperature regimes, four distinct seasons have been identified in India, viz., SW monsoon (June-September), NE monsoon (October-November), winter (December - February) and summer (March -May).

The important climatic elements used for characterization of agroclimatic resources are rainfall, temperature, sunshine, relative humidity, wind speed and potential evapo-transpiration (PET) as they directly or indirectly influence the plant suitability, adaptability, growth and productivity in any given region. These are described as under:

2.2.1 Rainfall

The Indian sub-continent comes under the influence of the Asiatic monsoon, which provides the major quantum of rainfall to the country. Though, rainfall of different amounts are received throughout the year, the rainfall received during monsoon season is the most important of all the weather elements as the country's agriculture production in rainfed agriculture is dependent (solely) on its quantum and distribution. The average annual rainfall of the country is 1200 mm which amounts to 400 Mha meter of water over the country's geographical area. About 80 per cent of the annual rainfall of the country as a whole and in certain areas 90 per cent of it is received during the SW monsoon period alone. Though the advancement of monsoon rains across the country from the Kerala coast to Jammu and Kashmir region is a regular phenomenon, annual variability with respect to onset and its distribution is noticed in all parts of the country.

Normally the onset of monsoon occurs on the first of June over Kerala coast and advances steadily further into different parts of the country. By the end of June, it covers the whole country except the extreme parts of western Rajasthan where it enters by mid-July. Thus, barring Kashmir in the extreme north and parts of Tamil Nadu, north and south interior Karnataka, Kerala and Bay island regions, most of the annual rainfall over the country occurs during SW monsoon season. The withdrawal of monsoon (the termination of main rainy season) begins from western Rajasthan by 1st October. The withdrawal process is slow to the east of 80°E over the Gangetic valley, NE India, eastern Madhya Pradesh, Vidarbha and other adjoining areas of Maharashtra and interior Peninsula. The NE monsoon rains merge with the withdrawing SW monsoon along coastal Andhra Pradesh, Tamil Nadu and Kerala. There is no considerable variability in the timing of onset and withdrawal of NE monsoon also.

The average SW monsoon seasonal rainfall of the country is about 850$\pm$90 mm. Orissa, east Madhya Pradesh, West Bengal and states in NE India, west coast and ghats, however, receive more than 1000 mm during the season. Also the sub-montane region extending from north Bihar to Jammu receives more than 1000 mm during the season. Among the heavy rainfall regions, geography influences the monsoon flow in receiving higher rainfall whereas other regions in eastern India receive heavy rainfall due to frequent passage of monsoon disturbances. The Peninsular India, south of 15°N receives less than 500 mm of rainfall during the SW monsoon season. In western and north-western India, the 500 mm isohyet covers, Saurashtra and Kutch to Delhi and then to Punjab, Haryana and Himachal Pradesh. Rainfall decreases rapidly to less than 100 mm in the extreme parts of western Rajasthan. Besides the monsoon rainfall, different regions of the country receive rainfall through some major weather systems viz., cyclones during SW and NE monsoon periods, and western disturbances during winter and pre-monsoon showers during summer months. Western disturbances provide good supplemental moisture to the winter season (*rabi)* crops grown on conserved soil moisture in north west India and are crucial for achieving higher productivity from these crops. Classification of rainfall zones in India is given in table 2.1.

Table 2.1: Classification of Rainfall Zones in India

Rainfall (mm)	Zone	Net sown area (%) (Approx)	Rainfall
< 500	Arid	16	Very low
500 to 750	Semi arid	14	Low
750 to 1100	Dry sub-humid	35	Medium
1100 to 1400	Moist sub-humid	24	High
>1400	Humid	8	Very high

2.2.2 Temperature

Next to rainfall (moisture regime), thermal regime is very important for active growth of the crop plants. Fortunately, most of the country's cultivable area is under tropical climate where temperature is not a limiting factor. The mean monthly maximum temperatures recorded in different meteorological sub-divisions show that the hottest month is May and the mean maximum temperature during summer varies from 30 to 40°C over most of the country. However, maximum temperature on individual days can reach above 40°C in many of the regions and even touch as high as 48 to 50°C in extreme arid parts of western Rajasthan. Among the northern most parts of the country, average temperatures remain below 28°C in Jammu and Kashmir in the hottest month, while they are 30°C in Sub-mountane region. The daytime temperatures decrease sharply with the onset of monsoon over the country and the fall in temperature is around 5 to 10°C with the onset of monsoon in June, the hot region shifts towards northwest India. The southern Peninsula (Tamil Nadu) also records high temperature during July and August due to low monsoon activity. With the withdrawal of monsoon, a slight increase in temperature of 1 to 2°C is generally noticed in many parts of the county. The cool season begins by early December. The minimum temperatures during the winter varies between (-)8 to 6°C in Jammu and Kashmir, 3 to 10°C in Indogangetic plains and northern Madhya Pradesh, 10 -15°C in central parts of Deccan Plateau. Minimum temperature sometimes falls below freezing point in the northern parts of the country (Punjab, Haryana, Himachal Pradesh and foot hills of Himalayas, etc.) and occurrence of frost, affecting agriculture crops is common, especially after the passing of western disturbances, which bring cool air in their wake.

2.2.3 Sunshine

Solar radiation affects a number of plant functions and regulates the plant growth and development. The mean actual daily duration of bright sunshine hours during different seasons recorded at various regions/places in the country show that sunshine is not a limiting factor during greater part of the year at majority of these places inspite of good cloud cover during monsoon season. It is observed that actual sunshine duration in the peak monsoon month (July and August) vary from 20 to 48 per cent of the possible sunshine in different parts of the country. Low periods of bright sunshine hours are observed in NE region, Gengetic West Bengal and Konkan region as these regions experience continuous cloudiness during monsoon season. Even during winter season when clear sky conditions are generally expected to prevail over the county, on an average only 50 per cent of the possible hours of sunshine are available in the NE region of the country. However, the percentage of actual to the possible sunshine hours over the rest of the country is about 70-90 per cent during this period. Thus, intercrops are greatly benefited as prolonged periods of sunshine/radiation is available for photosynthesis. The percentage of actual to that of possible sunshine hours during the peak summer month (May) is between 70-80 per cent, except in the NE region. After the withdrawal of monsoon, there is an increase in the percentage of actual to the possible sunshine in areas north of Deccan Plateau, whereas the values are low in Tamil Nadu, coastal Andhra Pradesh and Kerala which experience clouding under the influence of NE monsoon.

2.2.4 Relative Humidity (RH)

The values of relative humidity are indicative of the level of atmospheric saturation and evaporative demand of the air and play a crucial role in determining the water requirements of various crops. Also, high humidity coupled with favorable temperature periods provides a congenial environment for the development of pests and diseases. Relative humidity conditions experienced across different seasons indicate that in January (winter) the humidity ranges between 30-60 per cent in Gujarat, Rajasthan, madhya Maharashtra, Vidarbha; 40-70 per cent in Rayalaseema, Telangana, interior Karnataka, west Madhya Pradesh, east Madhya Pradesh; 60-80 per cent in Indo-Gangetic plains of Bihar, eastern Madhya Pradesh, parts of West Bengal, interior parts of Orissa and Tamil Nadu. High values of humidity (above 80%) are observed in coastal districts adjoining Bay of Bengal and Arabian sea and states of the NE region. During April (summer), low RH values ranging from 20-40 per cent are generally recorded in central and western parts of the country. On individual days in arid western Rajasthan, the afternoon values of RH can go below 5 per cent indicating the dryness of the atmosphere. With approaching monsoon, the RH values in June ranges between 30-60 per cent in most parts of the country. However, high RH (greater than 35%) are observed along the coastal districts and also in the NE region of the country. With the advancement of monsoon over the entire country (July), RH values increase at almost all the places and they ranges between 50-80 per cent. However, in the southern Peninsula, where monsoon activity is minimum during this period the RH values remain between 40-60 per cent. The RH values decrease with the withdrawal of monsoon current and during October, the value remain between 30-60 per cent in Rajasthan and Gujarat, 40-60 per cent in west Uttar Pradesh, west Madhya Pradesh and Punjab, and 60-70 per cent over the rest of the country.

2.2.5 Wind Speed

Wind regime influences the evaporative demand of crop plant to a great extent. Besides, stronger wind regime may lead to plant injury and destruction. The mean wind speeds are lowest during the winter months (3-6 kmph) and the wind regime builds up with the initiation of summer season and is at peak during the pre-monsoon periods of April to June (6-10 kmph). Stronger wind regimes are observed in western Rajasthan, Gujarat and parts of Tamil Nadu resulting from the surface configurations and thermal regime influences in these regions. Wind speeds, however, can be extremely high for shorter periods during intense cyclonic activity, resulting from the atsmospheric pressure gradients causing damage to standing crops and vegetation.

2.2.6 Potential Evapo-transpiration (PET)

Potential evapo-transpiration has become the most important and fundamental factor for characterization of agro-climates. The usefulness of this parameter in other fields such as water balance, irrigation assessment and scheduling, hydrology, etc. need not be over-emphasized. In modern concept annual potential evapo-transpiration exceeds precipitation, it is said dryland/

rainfed agriculture, there is no consideration for amount of rainfall. Therefore, this parameter is very important in rainfed agriculture. The annual PET ranges between 1400-1800 mm over most parts of the country. Parts of Deccan Plateau, Western Rajasthan and parts of Gujarat experience high PET of above 1800 mm. Peak values of PET are recorded in summer. During the SW monsoon season, PET values ranges between 400-600 mm in most parts of the country, except for the extreme south-eastern parts of the Peninsula where they are slightly higher. PET in the NE monsoon season and winter is generally low except along the west coast and over the Peninsula and decreases gradually northwards.

♦

3

DROUGHT

Low rainfall or failure of monsoon rains is a recurring feature in India. This has been responsible for droughts and famines. The word drought, generally, denotes scarcity of water in a region. Though, aridity and drought are due to insufficient water, aridity is a permanent climatic feature and is the culmination of a number of long term processes. However, drought is a temporary condition that occurs for a short period due to deficient precipitation for vegetation, river flow, water supply and human consumption. Drought is due to anomaly in atmospheric circulation.

Drought is a climatic anomaly, characterized by deficit supply of moisture resulting either from sub-normal rainfall, uneven distribution, higher water need or a combination of all these factors. Droughts lead to problems like wide spread crop failure, unreplenished ground water resources, depletion in lakes/ reservoirs, shortage of drinking water, reduced fodder availability, etc.

One of the sectors where immediate impact of drought is felt is agriculture. With increased intensity or extended duration of drought prevalence, a significant fall in food production is often noticed. Drought results in crop losses of different magnitude depending on their geographic incidence, intensity and duration. Drought not only affects food production at farm level but also national economy and overall food security as well. History of drought occurred and severe drought years in India is given in tables 3.1 & 3.2 as under:

Table 3.1: History of Droughts in India.

Period	Drought year	Number of droughts
1801-1830	1801, 1804, 1806, 1812, 1819, 1825	6
1831-1860	1832, 1833, 1837, 1853, 1860	5
1861-1890	1862, 1866, 1868, 1873, 1877, 1883	6
1891-1920	1891, 1897, 1899, 1901, 1904, 1905, 1907, 1911, 1918, 1920	10
1921-1950	1939, 1941	2
1951-1980	1951, 1965, 1966, 1971, 1972, 1974, 1979	7
1982-2015	1982, 1987, 2002, 2009, 2014, 2015	6

Table 3.2: Severe Drought Years in India.

Year	% area affected	Deficit of rainfall in affected area (%)
1918	71	49
1965	41	36
1972	47	35
1979	45	38
1987	50	45
2009	47	41

3.1 Definition of Drought

There is no universally accepted definition of **drought.** Early workers defined drought as prolonged period without rainfall. According to Ramdas (1960), drought is a situation when the actual seasonal rainfall is deficient by more than twice the mean deviation. American Meteorological Society defined drought as a period of abnormally dry weather sufficiently prolonged for lack of water to cause a severe hydrological imbalance in the area affected.

In general, drought means different things to different people. To a meteorologist it is the absence of rain while to the agriculturist it is the deficiency of soil moisture in the crop root zone to support optimum crop growth and productivity. To the hydrologist it is the lowering of water levels in lakes, reservoirs, etc, while for the city management it may mean the shortage of drinking water availability. Thus it is unrealistic to except a universal definition of drought for all fields of activities.

Prolonged deficiency of soil moisture adversely affects crop growth indicating incidence of agriculture drought. It is the result of imbalance between soil moisture and evapo-transpiration needs of an area over a fairly long period as to cause damage to standing crops and to reduce the yields.

3.2 Types of Drought

Drought can be classified based on duration and nature of users. In both the classifications, demarcation between the two is not well defined and many a time overlapping of the cause and effect of one on the other is seen. Thornthwaite (1947) classified drought into following four categories based on duration:

3.2.1 Drought Classified Based on Duration

(i) Permanent Drought: It is characteristic of the desert climate where sparse vegetation growing is adapted to drought and agriculture is possible only by irrigation during entire crop season. Permanent drought are associated with inadequacy of moisture/rainfall to meet the water requirement of crops during most of the years.

(ii) Seasonal Drought: It occurs in climates which have well defined rainy and dry seasons. The natural vegetation is made up of plants that produce seeds during the rainy season and then die plants that remain alive but becomes dormant in dry season. Most of the arid and semiarid zones fall in this category. Duration of the crop varieties and planting dates should be such that the growing season should fall within rainy season followed by the period having residual moisture storage.

(iii) Contingent Drought: This involves an abnormal failure of rainfall. It may occur almost anywhere especially in most parts of humid or sub-humid climates. It is usually brief, irregular and generally affects only a small area.

(iv) Invisible Drought: It can occur even when there is frequent rain in an area. When rainfall is inadequate to meet the evapo-transpiration losses, the result is borderline water deficiency in soil resulting in less than optimum crop yields. This occurs usually in humid regions.

3.2.2 Drought Classified Based on Relevance to the Users

(i) Meteorological Drought: In India, the definition for meteorological drought adopted by IMD is a situation when the deficiency of rainfall at a meteorological sub-division level is 25 per cent more of the long - term average (LTA) of that sub-division for a given period. Drought is considered; **moderate** if the deficiency is between 26 and 50 per cent and **severe** if it is more than 50 per cent.

In India, a year is considered to be a **drought year** in case the area affected by moderate and severe drought, either individually or together, is 20 to 40 per cent of the total area of the country and seasonal rainfall deficiency during southwest monsoon season for the country as a whole is at least 10 per cent or more. When the spatial coverage of drought is more than 40 per cent, it will be called as **all India severe drought year.**

(ii) Atmospheric Drought: It is due to low air humidity, frequently accompanied by hot dry winds. It may occur even under conditions of adequate available soil moisture. Plants growing under favorable soil moisture regime are usually susceptible to atmospheric drought.

(iii) Hydrological Drought: Meteorological drought, when prolonged, results in hydrological drought with depletion of surface water and consequent drying of reservoirs, tanks, etc. This is based on water balance and how it affects irrigation as a whole for bringing crops to maturity.

(iv) Agricultural Drought: It is the result of soil moisture stress due to imbalance between available soil moisture and evapo-transpiration of a crop. It is usually gradual and progressive. Plants can therefore, adjust at least partly, to the increased soil moisture stress. This situation arises as a consequence of scant precipitation or its uneven distribution both in space and time. It is also usually referred as **soil drought.**

When soil moisture and rainfall are inadequate during crop growing season to support healthy crop growth to maturity, this situation causes extreme crop stress and wilting is called **agricultural drought.** It is defined as a period of four consecutive weeks (of severe meteorological drought) with a rainfall deficiency of more than 50 per cent of the long term average (LTA) or with a weekly rainfall of 5 cm or less during the period from mid-May to mid-October (*kharif*) when 80 per cent of the country's total crop is planted, or six such consecutive weeks during the rest of the year.

(v) Physiological Drought: It is caused by some physiological changes in the plants when their roots fail to absorb water and nutrients from the soil due to increased concentration of soil solution. Presence of soluble salts in soil water increases the concentration of the solution and hence availability of water to plants is reduced although water remains in the soil. Similarly presence of certain chemicals in the soil interferes with the water absorption by the plants viz., humic acid restricts the water absorption in bog soils and the term is used as physiological dryness of the soil. The absorption of water is reduced in bog soils because of poor root growth of trees, which also happens due to more CO_2 and lack of oxygen.

3.2.3 Based on Time of Occurrence

(i) Early Season Drought: It occurs due to delay in onset of monsoon or due to long dry spells soon after the onset. It results in seedling mortality requiring re-sowing.

(ii) Mid Season Drought: Occurs due to long gaps between two successive rains and stored moisture becoming insufficient during the long dry spell. Its impact depends on crop growth stage, duration and intensity of the drought spell.

(iii) Late Season/ Terminal Drought: Occurs due to early cessation of rainfall and crops water stress at maturity stage, mainly during the years with late commencement or weak monsoon activity. Late season droughts are more critical as the crops yield is strongly related to water availability during the reproductive stage.

3.3 Development of Moisture Stress or Drought

Water deficits occur in the plant wherever transpiration exceeds absorption. It may be due to reduced absorption of water from soil, excessive water loss or both. After rain or irrigation, water deficits develop gradually and only for a very short period in the beginning. Though there is sufficient amount of soil moisture, water deficits develop due to higher transpiration than absorption especially on hot mid days. This temporary wilting is known as midday depression or **incipient wilting.** As days progress, the soil moisture is depleted and plants cannot absorb sufficient amount of water to meet the high demand. Water deficits therefore, occur for prolonged period. Under such conditions, stomata are open for a short period in the morning or evening and partially closed during the rest of the day. When the soil moisture reaches about (-)15 bars or (-)1.5 megapascal (MPa), plants show wilting symptoms most of the day but do not die and it is known as **permanent wilting** point. Plants recover when rain is received or irrigation is given. If the soil moisture is further allowed to deplete to a level of (-) 60 bars or (-) 6 megapascal (MPa) plants die permanently and this is known as **ultimate wilting** point.

3.4 Effect of Water Deficit /Drought on Physico – morphological Characteristics of Plant

Since water plays vital role in plant, it is not surprising that reduced water absorption and dehydration can have a deleterious effect on morphology and physiology of plants and ultimately the yield of the crop. A brief about the water deficit effect on various vital plant processes is given as under:

(i) Water Relations: Moisture stress may influence absorption, translocation and transpiration process in the plant. Actually water stressed plants manifest typical reduction in leaf water potential and relative water content. In drought conditions progressive and continuous decline in turgor of the plants is observed. At first absorption of water lags behind transpiration results in loss of turgor which appears due to rapid rise of transpiration as a result of increased dryness in the atmosphere. Water stress also cause increase in leaf and canopy temperature. As moisture stress increases, turgidity of guard cells also decreases. The stomata gradually close down reducing transpiration and its cooling action on the leaves. The increase in leaf temperature may be upto 8°C over ambient temperature.

(ii) Root and Shoot Growth: A little is known about the effects of drought on root development, although it has been suggested that root development is enhanced relative to shoot development during the following drought stress. Most research has shown a progressive reduction in rate of root elongation as drought is imposed, and in some cases root elongation ceases before shoot growth. In addition, as rates of root elongation are reduced the rate of suberization exceeds the rate of root elongation. The non suberized zone is reduced until it is virtually eliminated in non elongated roots. Such response to severe drought stress greatly reduce the water absorbing ability of roots.

(iii) Cell Division and Enlargement: Cell division and enlargement are very sensitive to drought stress. However, it has been observed that leaves of plants which have been exposed to drought stresse may contain a similar number of cells as leaves of non–stressed plants. Cell division may continue during stress at a reduced rate but upon removal of the drought stress a relatively rapid resumption of expansion growth or cell enlargement may occur. Nevertheless, during drought stress cell enlargement is affected and is the primary cause of plant stunting commonly observed under field conditions. Stem elongation or leaf enlargement can be inhibited by small diurnal water deficits that occur even with well watered plants on days of high radiation incidence. The effect of inhibited leaf enlargement is a reduction in the size of the photo–synthesizing surface causing a reduced crop growth.

(iv) Mineral Uptake: Since dry matter production is considerably reduced under dryland conditions, the uptake of NPK is reduced. Decreasing water availability under drought generally results in limited total nutrient uptake. An important effect of water deficit is on the absorption of nutrients by the root and their transport to shoots. Lowered absorption of the inorganic nutrients can result from interference in nutrient uptake and the unloading mechanism, and reduced transpirational flow. However, plant species and genotypes of a species may vary in their response to mineral uptake under water stress. In general, moisture stress induces an increase in N, a definitive decline in P and no definitive effects on K. As nutrient and water requirement are closely related, fertilizer application is likely to increase the efficiency of crops in utilizing available water.

Moisture stress also affects fixation, uptake and assimilation of nitrogen. Nitrogen fixation by leguminous plants is reduced by moisture stress due to reduction in legheamoglobin in nodules, specific nodule activity and number of nodules. There is negative correlation between specific nodule activity and stomatal resistance, nitrogenase activity in the nodules, which indicates the specific nodule activity, is less in stressed. Moisture stress also delays nodule formation in leguminous crops, but number of bacteroids containing cells are not affected by moisture stress mainly due to the reduction in nitrate reductase activity. Nitrogen deficiency increases stomatal resistance. Phosphorus deficiency increases the sensitivity of the most recently expanded leaves to moisture stress and stomata close while there is still sufficient turgor in mesophyl cells.

Mineral uptake is frequently reduced in drought stressed plants. One of the earliest sings of drought stress is the translocation of phosphorus from older leaves to the stems and meristematic tissues. The translocation of nitrogen closely follows that of phosphorus suggesting the occurrence of protein hydrolysis and alteration of normal cell function.

(v) Photosynthesis: Photosynthesis is reduced under water stress conditions due to stomatal closure, reduced supply of CO_2 and photochemical reactions. The gaseous exchange through leaves is governed by boundary layer (thin layer of still air on leaf surface) stomatal and mesophyll resistance. As water deficit increases, stomata start closing as a mechanism to reduce transpiritonal losses. As a consequence, the entry of CO_2 is also reduced. The conductance of mesophyll cell decreases with age of leaf. However, the decrease in conduction of mesophyll cells occurs earlier in water stress conditions. The decrease in mesophyll conductance of CO_2 also adversely affects photosynthetic process. During drought period, stomatal conductance and net photosynthesis decreases with increased water deficit. In susceptible plants usually net photosynthesis is completely inhibited by about 14 days after start of drought stress.

Direct and indirect effects of drought on rate of photosynthesis and finally on carbohydrate reserve of the plant is reported by several scientists. Insufficient supply of CO_2 to plants and dehydration results in closure of stomatal openings and lower the net photosynthesis in drought affected plants and reduced starch level in plants is reported.

(vi) Respiration: The influence of drought stress on the respiration of plants has been presented in several reviews and it is generally agreed that under severe water stress respiration is drastically reduced. Nevertheless, the influence of limited stress on respiration is debatable because some investigators have reported an initial stimulation in rates of respiration at low levels stress whereas others have reported a continuous reduction in respiration with stress. Membranes are the site of mitochondrial respiration and a structure - function relationship exists between membrane and respiration, changes in the phospholipid or protein structure of the mitochondrial membrane may cause the suppression of oxidation rates of succinate, malate-pyruvate, and exogenous nicotinamide adenine dinucleotide (NADH) in water stressed tissues. Therefore, the ultimate cause of the inhibited oxidation of respiratory substrates measured in mitochondria isolated from drought stressed corn may be due to altered membrane structure caused by depletion of water.

(vii) Protein Synthesis and Proteolysis: Interruption of protein synthesis and proteolysis generally occur under drought conditions. There is a close relationship between protein synthesis and ribonucleic acid (RNA) levels. Drought stress has been shown to impair the nucleic acid system intimately correlated with protein synthesis. Some other scientists also reported that a decline in protein synthesis in drought affected leaves is due to a deficiency of cytokinins. Addition of cytokinins to leaves of a number of annual plants has been shown to retard chlorophyll degradation and has promoted amino acid incorporation and protein synthesis. Nevertheless, it has been suggested that the protein breakdown injures the drought stressed plants due to the accumulation of a toxic product of protein breakdown, such as ammonia, rather than due to a protein deficiency.

High levels of free amino acids and amides have been associated with drought stress in plants. Proline has been especially high in stressed plants although understanding for this response is unknown. It has been suggested that free proline serves as a storage compound for carbon and nitrogen during drought, when both starch and protein syntheses are inhibited.

It is reported that an interruption in protein synthesis and proteolysis which is directly related with ribonucleic acid (RNA) levels of the plants, lowers protein synthesis rate resulting in lower nucleic acid due to water stress. It was concluded that lower protein synthesis in drought affected plants is due to deficiency in cytokinins. High level of free amino acids and amides have been associated with drought stress in plants and proline, especially has been found to be very high.

(viii) Hormonal Relationship: As a consequence of water deficits, hormonal balance is altered. The activity of growth promoting hormones like cytokinin, gibberellic acid and indole acetic acid decreases and growth regulating hormones like abscisic acid, ethylene, betain, etc., increases. The translocation of growth promoting hormones is also reduced by moisture stress. Abscisic acid content is inversely related to leaf water potential. Abscisic acid acts as water deficit censor to minimize the loss of tissue water potential. It controls stomata and thus reduces the water loss when there is moisture stress. Ethylene production induced by moisture stress is considered to be the cause for leaf and fruit drop. Betain is another hormone produced by the moisture stressed plants and it is used as an indicator of moisture stress.

(ix) Protoplasmic Dehydration: As the tissues become desiccated, protoplasm become increasingly dense and its viscosity gradually increases. When dehydration is severe, it becomes rigid and brittle.

(x) Grain Yield: Three important stages of crop growth in relation to drought stress should be considered. These are stages of floral initiation and inflorescence development, anthesis and fertilization, and grain filling. Good evidence exists that a slight drought stress can reduce the rate of appearance of floral primordial. However, removal of a slight drought stress in barley has been shown to cause a more rapid rate of primordial initiation than for non-stressed plants. When a severe drought stress was imposed, the recovery from drought was unsatisfactory and the total spikelet number was greatly reduced. Primordial initiation is more affected by drought stress than spikelet development, and thus, stress at the former stage can alter grain number more than at the latter stage.

Drought stress at anthesis and fertilization will reduce the number of kernels because of the dehydration of pollen grains. Another factor is that pollen grain germination and pollen tube growth down the style into the ovary and ovule are affected. Wilting of the styles interferes with pollen tube growth, and one might expect such a response to be more pronounced in species possessing long styles, such as corn. Crop plants which shed pollen over an extended period of time will be more likely to avoid the influence of drought at this stage of growth than crop plants which shed pollen in a relatively short period of time.

Drought stress at the stage of grain filling is pronounced as yield development, expressed by weight per grain, requires the accumulation of photosynthate in the grain. The two sources for the accumulation of these assimilates are photosynthesis in the grain itself and translocation from other plant parts. A portion of the photosynthate is synthesized prior to anthesis, stored in the stem or other organs and later translocated to the developing grains, but the greatest accumulation is synthesized after anthesis. Drought stress has been shown to reduce translocation from the leaves, and as drought hastens maturation, this response in addition to reduced photosynthesis contributes to lower grain yield.

3.5 Crop Adaptations to Drought

The ability of crop to grow satisfactorily under water stress is called drought adaptation. Adaptation is structural or functional modification in plants to survive and reproduce in a particular environment. The mechanism of plant adaptation to drought is given in fig. 3.1 as under:

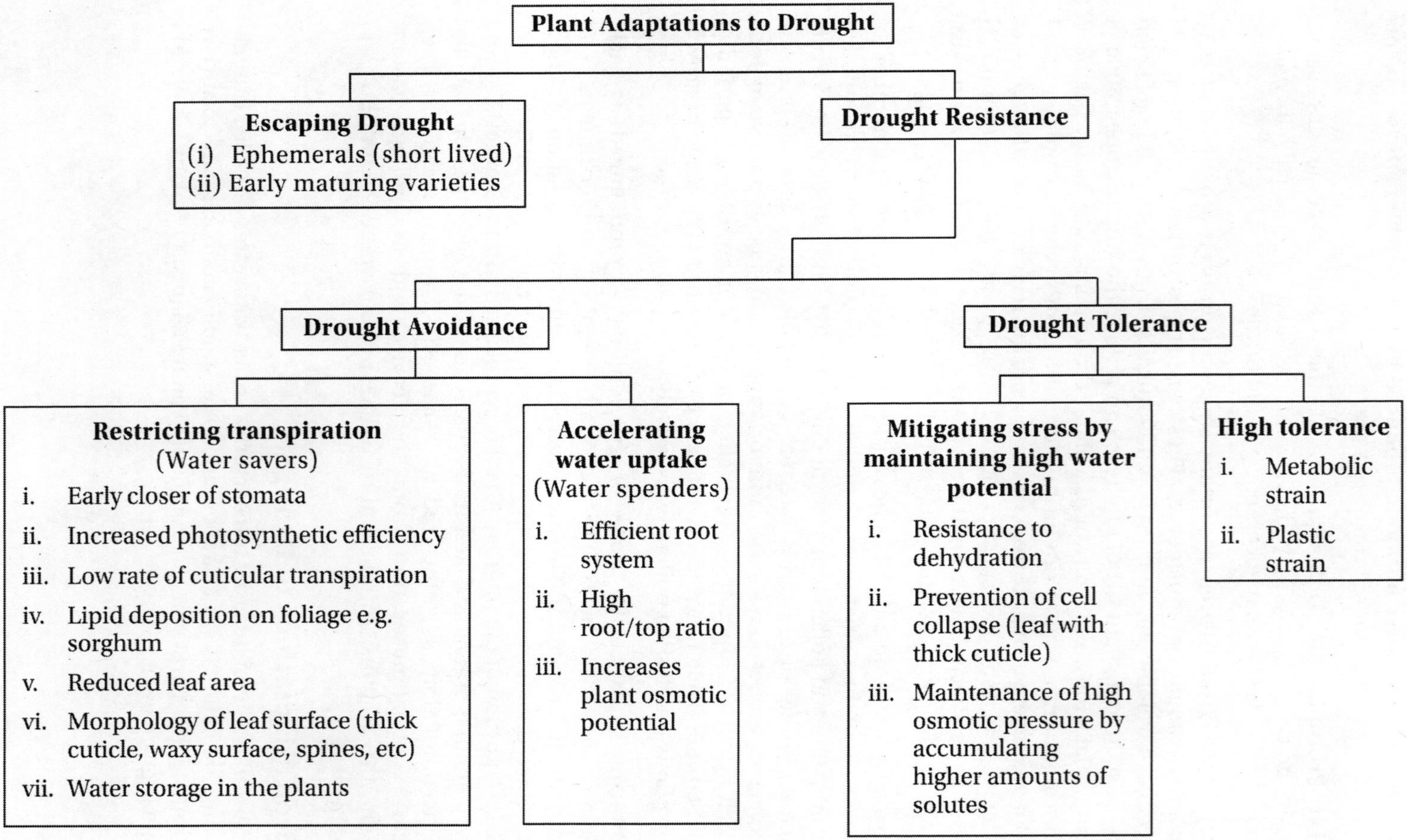

Fig. 3.1: Plant Adaptations to Drought.

3.5.1 Escaping Drought

Evading the period of drought is the simplest means of adaptation of plants to dry conditions. Many short duration desert plants, the so called **ephemerals,** germinate with the rains and have an extremely short life period of 5 to 6 weeks which is confined to the rains period. These plants have no mechanism for overcoming moisture stress and are therefore, not drought resistant.

In cultivated crops, the ability of a cultivar to mature before the soil dries is the main adaptation to drought in dry regions. However, only very few crops have such a short growing season to be called as ephemerals. Certain varieties of pulses like greengram, blackgram maturing in 60 days duration can be included in this category. In addition to earliness, they need drought resistance because there may be dry spells within the crop growth period of 60 days adaptations to moisture stress.

3.5.2 Drought Resistance

Plants can adapt to drought conditions in two ways: avoiding stress and tolerating stress.

3.5.2.1 Drought Avoidance

It is the ability to maintain favourable water balance and turgidity even when subjected to drought thereby avoiding stress and its consequences. Favourable water balance can be achieved either through conserving water by **restricting transpiration** (water savers) or by **accelerating water uptake** (water spenders). The mechanisms for conserving water are regulating stomatal opening, increased photosynthetic efficiency, low rates of cuticular respiration, decreasing transpiration by lipid deposition on leaves, reducing leaf area, stomatal frequency and location and presence of awns. Water uptake can be accelerated by efficient root system, high root to top ratio, differential osmotic potential of plants and change of water spenders to water savers. The mechanisms of conserving water or accelerating water uptake in plants is described as under–

(a) **Mechanism for Conserving Water in Plants:** The drought affected plants develop certain changes in the anatomical characteristics of plant for an effective conservation of the cell sap against accelerated rate of transpiration due to rise in plant body temperature. These changes can be described as under:

(i) **Stomatal Mechanism:** In certain plant types mostly in cereals, it has been found that they open their stomata only for few hours early in the morning when moisture is at its minimum and they close the stomata for rest of the day hours which result in an efficient photosynthesis with least loss of water. These type of plants survive for more longer period under the drought conditions than those which open their stomata late in the morning and keep them opened for rest of the day hours that favours more water loss due to respiration.

The stomatal mechanism of opening and closing differs greatly from crop to crop or even within the varieties of the same crop. However, the mechanism of closing of stomata or stomatal regulation of water loss is not operative beyond a certain thresh hold level of water content in plants and beyond this level of water the closure of stomata would lead to reduced photosynthesis which may result into drought induced starvation injury and subsequently in death of the plant.

(ii) **Increased Photosynthetic Efficiency:** Closure of stomata to reduce the transpiration loss of water, results in a lower rate of photosynthesis and best possibility to overcome this problem is an increased CO_2 assimilation by plant for given stomatal opening. The CO_2 assimilation capacity of C-4 group of plants to fix most of the CO_2 into the form of organic acids. The C-4 species have higher growth potentials and more efficient use of water as compared with plants of Calvin cycle or C-3 plants under the same moisture regime. The carboxylating enzyme, which has a high potential activity and a high affinity for CO_2 helps in a high rate of CO_2 fixation and maintenance at relatively restricted stomatal opening. These physiological changes use to inhibit the photo-respiration and thereby the net photosynthesis is enhanced.

(iii) **Low Rate of Cuticular Transpiration:** Plants use to develop cutine, a waxy coating over cuticles for reducing the cuticular transpiration of water. The intensity of cutine covering depends upon the severity of the drought. The dwarf wheat varieties of durum group, triticale, barley use to get the waxy coating under rainfed conditions to survive under drought. Thus, some plants have lower rate of cuticular transpiration (viz, xerophytic plants such as cactus) by which they conserve water and survive in drought conditions for a pretty long time.

(iv) **Decreasing Transpiration:** The transpirational loss of water through other plant parts is a very significant loss. It is reported that when soybean is exposed to moderate light especially under drought the plant produce lipid on the surface and provide a covering through which the transpiration of water is reduced to minimum. Thus, with limited supply of water plants grow better and produce good yield.

(v) **Reduced Leaf Area:** The water loss through transpiration is directly proportionate to the leaf area and plants can survive longer, if they can reduce their surface leaf area. The xeromorphic plants reduce their transpiring surface and find fit for drought conditions. In other plants the rolling and curling of leaves at noon time, especially under dry conditions reduce the transpiritonal loss of water to the extent of 55 per cent in arid zones and 75 per cent in desert areas. These leaves take their normal shape in the evening when transpiration is at its minimum, thus the plants escape the heat of the day.

(vi) **Leaf Surface:** Various morphological adaptations of leaves, help to reduce the rate of transpirational loss of water for prolonged survival of the plants under drought conditions. The most pronounced changes in the leaf surface are development of thick cuticle, deposition of more lipids and waxy covering development or spines, etc. Some plants convert their leaves into thick flattened leaves covered with spines such as cactus, etc. which have very low transpirational requirement of water for normal functioning.

(vii) **Position of Stomata and its Functioning:** Position of stomata present on the leaf surface or on slightly elevated position come in direct contact with hot air current and transpire more water for the protection of plants life from desiccation but stomata situated in cavities or in depressions use to retard the transpirational loss of water from the leaf surface. The duration of the stomatal opening also governs the extent of water transpired from the leaf surface. The plants affected by drought, use to bear less number of stomata which are located either in depressions (in cavities) and they retard the rate of transpiration of water.

(viii) **Effect of Awns:** Awned varieties of cereals like wheat and barley are the best adapted under drought conditions. Under drier conditions the awned varieties of wheat give better yield than the awnless and awns contribute to about 12 per cent to the total dry matter production of the drought affected plants. These awns posses chroloplasts and stomata and perform the additional phosynthetic activities resulting in production of yields. The drought affected plants undergo certain morphological changes like shorter leaves, production of waxy leaf surface, thick cuticular layers, development of awns, possessing less number of stomata situated in depression or in cavities of the leaf surface, etc. help in reducing the water loss through transpiration and plants live longer having better growth, development and yield.

(b) Mechanisms to Accelerate Water Uptake by Roots

If the plants bear well developed and extensive root system then they can withdraw water from relatively deeper layers of the soil but contrary to this the shallow rooted plants are often victim of drought. The mechanism of water uptake totally depends upon the root system of the plants which may be described as under:

(i) **Effective Root System:** The vertical growth of the roots is more helpful for an efficient adaptation of plants to dry conditions. Deeper roots with well developed better laterals and extensive root hairs are more beneficial to fight the drought. Thus, the density of roots per unit volume of soil and number of secondary hair roots are the proper index for finding out the drought resistance of plants, however, the water uptake from the soil is partly controlled by presence of salts and their concentrations in the soil solution.

(ii) **High Roots to Top Ratio:** Deeper the roots better is the drought resistance in plants. The water movement to deeper layers in soil results in more root elongation and hence the root growth rate exceeds considerably from that of the stem. This type of growth results in better uptake of water because of deeper roots and also in reduction of water loss through transpiration as volume of the top or shoot remains lower with lesser number of stomata and cuticles per plant.

(iii) **Differences in Osmotic Potential of the Plants:** The water movement in plants depends upon their osmotic pressure. Drought affected plants can withdraw water from soil even after wilting point, whereas, plants under irrigated conditions cannot do so. Therefore, plants grown under dry conditions use to have high osmotic pressure and they survive for relatively longer period.

(iv) **Development of Water Economy in Plants:** Plants use to transpire with normal speed under regular water supply but a restricted water supply or reduction in availability of water to plants results in a reduced rate of transpiration by plants. Thus, the water spenders become water savers to fight better with drought. This change comes automatically in the plants and a high drought resistance is developed in the plants.

3.5.2.2 Drought Tolerance

Plants can tolerate drought either by mitigating the actual stress or by showing high degree of tolerance to stress. **Mitigating the stress** by resistance to dehydration and by preventing leaf collapse, permit the plants to maintain a high internal water potential inspite of drought conditions. **Tolerating the stress** by resistance to metabolic strain (*starvation* acid protein loss) and plastic strain (increased resistance to stress due to exposure to sublethal stress for long period) can increase the plant ability to resist and survive under conditions of soil moisture stress. Physiological adaptations in plants to induce drought tolerance is described as under:

(a) **Mitigating Moisture Stress:** Some plants, when exposed to drought conditions, develop adaptation by maintaining high turgor or the internal water potential even under very high drought conditions. Thus by maintaining the turgor in these plants avoid secondary drought induced stress caused by dehydration. These drought affected plants behave in the following two ways for mitigating drought.

(i) **Resistance to Dehydration:** The morphological adaptations of drought affected plants help in accumulation of sufficient solutes which help in maintaining a higher cell turgor pressure that prevents dangerous internal desiccation and the plasma itself becomes more drought enduring. Thus, the plants keep on growing in normal way even under severe drought conditions.

(ii) **Prevention of Cell Collapse:** The drought induced morphological adaptation like thick cell wall, sunken stomata lipid or waxy coating over leaves, presence of spines, etc. prevent wilting or cell collapse by reducing the transpiritonal loss of water. With further fall in soil moisture status the period of stomatal opening is reduced and other morphological adaptations to reduce water loss become operative. Thus the leaf collapse is avoided and plants are saved.

(b) Tolerating Moisture Stress

The drought tolerance of a plant depends upon the ability of the stomatal guard cells to remain tugid even when the leaf cells are wilted, which is possible due to increase in the solution of the guard cells. The dehydration compensation point (it is water potential at which the volume of CO_2 evolved during respiration and CO_2 absorbed during photosynthesis becomes equal) is reduced with increase in drought tolerance of plants. At the dehydration compensation point the photosynthesis is completely stopped and subsequently the water loss is also reduced. Under this stage opening of stomata would be harmful. The plants which reduce their metabolic activity below dehydration compensation point would need lower carbohydrate and hence they would be more drought tolerant at lower net assimilation rates. These plants remain alive and do not starve suddenly. The disturbances of the water balance in the plant result in protein loss and the plants with more loss of protein die quickly than those which avoid loss of protein. The protein loss is avoided by drought resistant plants by two important processes like decreased rate of breakdown which is attained due to prevention in the loss of Ribonucleic acid (RNA) and the plants continue to synthesize the protein and secondly accumulation of more proline checks the protein breakdown and these plants become more drought tolerant and keep on growing under moisture stress conditions.

3.6 Managing Drought Stress

Drought stress effects can be managed by production of the most appropriate plant genotypes together with manipulation of agronomic practices like sowing time, seed hardening, plant density and geometry , soil and nutrient management, weed management, etc. To develop drought tolerant genotypes efforts have been made to produce drought tolerant genotypes using the knowledge of responses of plants to drought stress and mechanisms involved as discussed above. The most important strategies may include; selecting the desired materials as in traditional breeding using molecular and biotechnological means, including production of genetically modified or transgenic plants and inducing drought tolerance in otherwise susceptible plants by priming and hormonal application. An account of these efforts is elaborated below.

(a) Selection and Breeding Strategies

Conventional breeding has been based on empirical selection for yield. These include (i) increasing the uptake of available water (ii) improving bio-mass production per unit transpired water and (iii) partitioning of produced bio-mass towards the harvested product.

(b) Molecular and Functional Genomics Approaches

For more than two decades, molecular and biochemical studies have identified many of the abscisic acid and stress responsive genes and a few of the transpirational factors responsible for their induction in crop plants.

(c) Induction of Drought Resistance

Drought resistance can be induced by adopting various strategies. Of these, exogenous use of various growth regulating and other chemicals has proven worthwhile in producing drought resistance at various growth stages in a number of plants as follows.

(i) **Seed Priming:** One of the short term and most pragmatic approaches to overcome the drought stress effects is seed priming. Seed priming is a technique by which seeds are partially hydrated to a point where germination related metabolic processes begin but radical emergence does not occur.

(ii) **Use of Plant Growth Regulators:** Foliar application of plant growth regulators, both natural and synthetic, has proven worthwhile for improving growth against a variety of abiotic stresses. Drought stress alone inhibited increase in length and fresh weight of the hypocotyl, while applied levels of gibberelic acid reversed this effect. In this case, gibberellic acid partially increased the water status of the seedlings and partially sustained protein synthesis.

(iii) **Use of Osmoprotectants:** Osmoprotecatns are involved in signaling and regulating plant responses to multiple stresses, including reduced growth that may be part of the plants adaptation against stress. In plants, the common osmoprotectants are proline, trehalose, fructan, mennitol, glycine-betaine and others.

(iv) **Application of Silicon:** Silicon is the second most abundant element in soils and a mineral substrate for most of the worlds plant life. Ample evidence is available indicating that when silicon is readily available to plants, it plays a significant role in their growth, mineral nutrition, mechanical strength and resistance to several stresses and an important role of silicon in water transport and maintenance of root growth under drought stress.

Different crop management practices for drought management are discussed in chapter 6 of this book.

◆

4

SOIL AND WATER CONSERVATION TECHNIQUES

Soil and water conservation techniques/practices are important component in dry land agriculture and can play a vital role in erosion control and moisture conservation. These practices usually involve creating mechanical barriers across the direction of flow of water and thus retard or retain the runoff. They supplement agronomical practices when the latter alone are not adequate to handle the runoff. The important principles kept in view while planning for soil and water conservation measures are:

— Increase the time of concentration for allowing more runoff water to be absorbed and held by the soil.

— Break a long slope into several short ones for reducing the flow velocity below critical limits.

— Protection of drainage channels against damage owing to excessive runoff in them.

— Prevent excessive soil and water losses.

Soil and Water Conservation Techniques / Practices

4.1 Mechanical/ Engineering Techniques

4.1.1 Terrace Level Practices

(i) Contour Bunding: On arable lands, bunds are important supplement to contour farming and vegetative barriers on keylines. Bunds are simply embankment like structures, constructed across the land slope. When bunds are constructed on field boundaries without reference to contour they are called **field bunds.** Contour bunds are generally used in relatively low rainfall areas for the purpose to control soil erosion and to store the rain water. Contour bunds are constructed for the slope upto 6 per cent in low rainfall areas having the annual rainfall less than 600 mm and where soil moisture is limiting factor for crop production (Fig 4.1). In the contour bunds, cultivation is not possible on bund itself. This practice is suitable for all types of relatively permeable soils but not for clay or deep black cotton soils. These soils generally crack in hot weather and this is the main demerit for the bund construction. Different types of bunds constructed as per need are described as under–

Fig. 4.1: Contour Bunding

Types of Bunds

- **Side Bunds:** Bunds constructed at extreme end of contour bund, running along the slope are called side bunds.
- **Lateral Bunds:** Bunds constructed along the slope in between two side bunds in order to prevent concentration of water along one side and to break the length of contour bund into convenient bits are called lateral bunds.
- **Supplemental Bunds:** Bunds constructed between two contour bunds so as to limit a horizontal spacing are called supplemental bunds.
- **Marginal Bunds:** Bunds constructed along margin of the watershed, road margins, river of stream margins, gully margins, are called marginal bunds.
- **Graded Bunds:** Graded bunds or channel terraces are laid along a pre-determined longitudinal grade instead of along the contour. Graded bunds consist of constructing wide and relatively shallow channels across the slope very near to the contour ridges and at critical intervals. These terraces act as drainage channels for regulating the excess runoff from the field and draining it with non-erosive velocity. The purpose of graded bunding is to make runoff water trickle rather than to rush out.

  Graded bunds are constructed for the land having slope upto 6 per cent in medium to high rainfall areas receiving annual rainfall of more than 600 mm and in soils having poor permeability *i.e.* clayey soils.

  For the planning and designing of graded bunds the first item to be considered is the location of the most desirable terrace outlets. Preferably the outlets must be located in the natural waterways.

- **Grade:** Graded bunds may have uniform grade or variable grade. Uniform graded bunds are suitable where the length of bunds is less and discharge behind the bund is not much. Variable graded bunds are suitable where the length of bund and discharge are more. Variable grades are provided in the different sections of bund so that the velocity of flow is within non- erosive limit. A grade of 0.2 to 0.4 per cent is provided depending on soil type. Bunds of 0.8 m^2 cross sections at a vertical interval of 0.7 m with a channel (grade of 0.1 to 0.25%) on the upstream side connected to a grass waterways are advocated. In case of impervious soils, it may be started with 0.2 per cent at upper end and increase to 0.4 per cent at the outlet.

(ii) Puerto Rico Terrace (PRT): Puerto Rico Terrace of dry stone is constructed along the contours, which develops into level bench terraces due to shifting of soil down the slope every time after ploughing. This type of terrace is specially suitable for arable lands when slope is more than 6 per cent and where depth of soil is shallow. These are mostly constructed in the areas where stones are easily available at the site. For increasing crop productivity in rainfed areas these terraces are very effective (Fig. 4.2)

Fig. 4.2: Puerto Rico Terrace

In these terraces, stone bunds or Stone Wall Terracing (SWT) of size 0.60 m × 0.45 m are erected at field boundaries on both the sides of nalla/valley at the designed horizontal intervals across the slope.

(iii) Stone Wall Terracing (SWT): Stone wall terrace is a stone barrier placed across the small gully or cultivated valley. It is provided to impound water and cause sedimentation in an eroded field. The impounded water conserves moisture, which can be used for crops (Fig. 4.3).

In cultivated valleys of hilly tracts, stone wall terrace is commonly used as soil and water conservation measure, preferably at sites where stones are easily available in nearby areas. The cross section and spacing of stone wall terracing is decided by taking into account the slope of the land, rainfall parameters, heights, expected depth of flow and future prospects of increasing the height.

Fig. 4.3: Stone Wall Terracing

(iv) Bench Terracing: Bench terracing is the most widely adopted mechanical measure of soil and moisture conservation, suitable for hilly areas with a slope range of 6 to 33 per cent. Bench terrace consists of construction of step like fields along the contours by half cutting and half filling (Fig 4.4). In this original slope of the land is converted into level fields and thus all hazards of erosion are eliminated. It is one of the most popular mechanical soil conservation practice adopted by the farmers of hilly areas where sufficient soil depth is available. In the area where climatic conditions favour the growing of certain cash crops like potato, tea, coffee, etc., the hill slopes are to be bench terraced before the area can be put for cultivation of the crops. It is also adopted for converting sloping lands to irrigate fields or for orchard plantations.

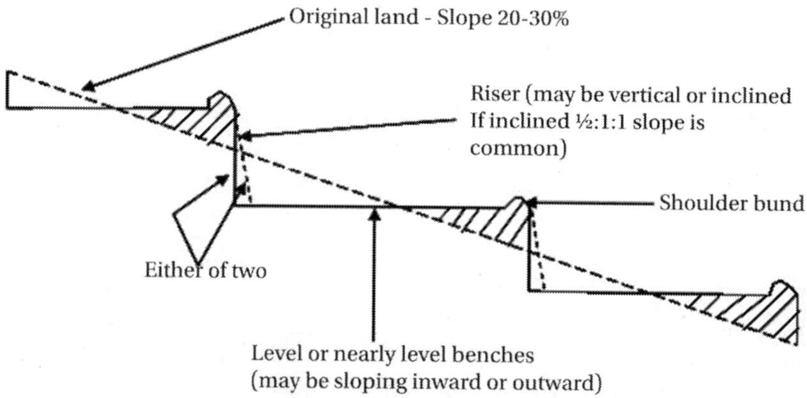

Fig. 4.4: Sketch of Bench Terracing Mechanism

Types of Bench Terraces:

- **Level or Table Top Bench Terrace:** Table top bench terraces are suitable for areas receiving medium rainfall which is evenly distributed and have highly permeable deep soils.
- **Sloping Outward Bench Terrace:** Bench terrace of sloping outwards are effective only in low rainfall areas with a permeable soil of medium depth.
- **Sloping Inward Bench Terrace:** This type of bench terrace are effective in heavy rainfall areas. These inward sloping bench terraces have a drain on inner side, which has grade along its length to convey the excess water.
- **Zing Conservation Terrace:** Zing conservation terracing is adopted for undertaking cropping programme in areas having rolling topography and on soils relatively heavy in nature ranging from sandy loam to clay loam with moderate depths. In medium depth soils having infiltration rate of more than 8 mm/hr, Zing conservation terrace of 1:3 levelled to unlevelled portion with a ridge section of 1.2 to 1.5 m^2 at a vertical interval of 1.0 to 1.5 m have been found suitable.

(v) Contour Border Strips: These strips are to be laid out in medium to deep black soils having infiltration rate more than 8 mm/hr. The strips are constructed along the contour key lines at a vertical interval of 0.3 m with a suitable drop structure at the end of each strip in the waterway with ridge section of 0.24 m^2 and stabilization of vegetation.

(vi) Graded Border Strips: These strips are formed in between the contour bunds or graded bunds. It is a small section bund having a cross section of 0.2 m^2. The interval of bunds vary from 10–15 m with the length of 100–200 m. It is useful for deep red soil areas where the land slope is 2–3 per cent.

4.1.2 Inter-terrace Management Practices:

(i) Tillage Practices: Off season tillage or early tillage after the harvest of standing crop or summer ploughing keeps the land in rugged condition and helps to avoid the runoff apart from having other benefits such as reduction in pests and weeds, burial of previous crop residues and reduction in lapse of time between first rain and actual seeding time. Shallow tillage soon after rain is beneficial for retaining water in the soil profile particularly in the fine textured soils. Similarly, deep ploughing with inversion plough besides creating surface inequalities forms an infinite number of small depressions among the clods as storage spaces for impounding and later leisurely absorbing rain water received form intense summer storms. Due to increase in opportunity time for infiltration and reduced peak runoff rate, runoff and soil loss are low.

(ii) Compartmental Bunding: In lands sloping up to 2 per cent, compartmental bunding is a best low cost *in-situ* water conservation measure. Compartmental bunds are formed in a field with a size of 40 m^2 with the dimensions of 8m × 5m. The longest bund is usually formed across the slope while shorter bunds along the slope. By this method the entire field is divided in to small sections, which helps in storing the initial rainfall, and permitting increased infiltration rate (Fig. 4.5). This method is recommended for medium and deep black soils where *rabi* cropping is taken up.

Fig. 4.5: Compartmental Bunding

(iii) Furrowing: Plough furrows opened at frequent distances intercept runoff and as a result more rain water is conserved in between furrows. Furrows may be at a distance ranging between 0.45 m and 10.0 m rolling lands, the spacing may vary from 2.5 m or 4 to 8 crop rows. The furrows be laid out across the major slope. The depth of furrows could be 20–30 cm. Furrowing needs to be done prior to sowing of crops. The furrows could be tied at certain distances by earthen check constructed at 3–5 m interval. The series of earthen dams hold rainwater *in-situ* and provide more opportunity time for it to infiltrate in to the soil.

(iv) Broad Furrows and Ridges: This system has proved effective on medium to deep black soils having a slope of 2 per cent or less. In this practice the land is laid out into broad furrows and ridges across the slope. Furrows hold more moisture and more moisture requiring crops may be grown in furrows. Ridges may be uncropped or planted with long season component crops like pigeonpea. In heavy rainfall areas in upland situation, such a system helps to grow paddy in the furrows and arable cotton on the ridges to make best use of available rains.

(v) Broad Bed and Furrow (BBF) System: The broad bed and furrow system is one of the most efficient land management practice evolved by International Crop Research Institute for Semi Arid Tropics, Hyderbad (ICRISAT) for *in-situ* moisture conservation in deep black soils. Primary tillage operations are carried out immediately after the harvest of the previous crop with the available soil moisture and the secondary tillage with the help of the summer and pre-monsoon showers. Land is levelled to give a uniform slope of 0.4 to 0.8 per cent. The broad bed and furrows are formed with the help of 'Tropiculator' a special equipment. The beds are 120 cm wide and furrows are 30 cm wide and 15 cm deep. The length of the furrow ranges from 100 to 200 m depending on the slope of the land (Fig. 4.6).

The broad-bed and furrow system of land preparation gives several advantages such as it conserves moisture, reduces erosion, drains off excess water to ensure surface drainage, weed control, reduces power inputs and provides graded furrows for crop drainage and supplemental irrigation.

Fig. 4.6: Broad Bed and Furrow (BBF) System

(vi) Scooping: It is a low cost *in-situ* conservation practice. The scoops or small depressions created with the help of a wooden hoe with plank attached to basin lister (scooper) holds water and allows it to infiltrate. This technique is mostly recommended for deep black soils where post rainy season cropping on stored soil moisture is predominant.

4.2 Mulching

Mulches are covering material applied on soil surface to check evaporation and improve soil moisture. Mulches improve soil water by reduction of evaporation, runoff and weeds and increase in infiltration. Evaporation occurs by absorbing energy for vaporization and escape of water molecules into the air. Under field conditions, energy is supplied by solar radiation and wind provides transport to vapour molecules. Application of mulches on the soil surface obstructs the solar radiation reaching the soil. It also checks the escape of water vapour by physical obstruction. The mulch present on the soil surface receives the energy of beating rain drops. Thus, it protects the breaking down of soil aggregates and subsequent sealing. As rain drops slowly trickle through the mulch, infiltration is more. Mulch slows flow velocity of runoff. As there is more opportune time, more water enters the soil and thus reduces runoff. Mulches also helps in moderation of soil temperature, improving soil fertility & soil structure and weed control.

4.2.1 Types of Mulches

Based on the materials used for mulching, mulches can be classified into following categories:

(i) Soil Mulch or Dust Mulch: Soil mulch is a thin layer of loose soil surface that can be created by frequently stirring the soil with surface tillage implements. Soil mulch of surface 5–8 cm dry soil (fine dust particles) effectively reduces the evaporation losses by obstructing the upward movement of soil moisture through capillary action. The soil mulch also prevents deep cracks in soil (especially vertisols) by reducing the direct action of atmosphere and hence evaporation is also reduced. After the preliminary loss of soil moisture, soil mulch effectively controls further evaporation from subsoil by breaking the capillary continuity. Among the different mulches, soil mulch is the cheapest. This type of mulching is highly suitable for moisture conservation in medium to heavy texture soils that tend to shrink and crack deeply. Such cracks when penetrate deep, may cause considerable loss of moisture from the root zone. These cracks are filled and covered by the loose layer of dust mulch. Its usefulness is doubtful in alfisols.

(ii) Straw and Stubble/Organic Mulch: Straw and other crop residues like stubbles, groundnut shells, cotton stalks, etc., can spread on soil surface, works as mulch for moisture conservation (Fig. 4.7). Straw mulches reduce both the amount of energy absorbed by the soil and its movement above the soil and hence reduce evaporation. However, the availability of adequate quantity of crop residues is a problem for use as mulches.

Fig. 4.7: Straw/ Stubble Mulch

(iii) Plastic Mulch: Plastic mulches are very effective as mulches for evaporation control, provided cost is not a limiting factor. The plastic mulches may be either white or black. Black plastic mulches will absorb the solar radiation and enhance the soil temperature for hastening the germination of winter crops like wheat, barely, etc. White plastic mulches will reflect the incident radiation and reduce evaporation of soil moisture. Polyethylene, polyvinyl chloride are used as mulching materials (Fig. 4.8).

Fig. 4.8: Plastic Mulch

(iv) Chemical Mulch: Chemicals like hexadecanol, a long chain alcohol when mixed with surface 5 mm of soil can reduce evaporation by about 40 per cent. The surface layer of a treated soil dries out more rapidly than that of untreated soil, creating a diffusional layer to evaporation.

(v) Vertical Mulching: Sub–soiling is probably the most effective methods of breaking hard pans to improve root penetration, aeration and water percolation. Unfortunately, the effects of sub-soiling are not long lasting. The slots left by the sub-soiler are closed and sealed off within few months. In order to prolong the beneficial effect of sub-soiling, a method called vertical mulching has been developed. The object of vertical mulching is to fill slots with organic matter and keeping them open and functional for a longer period. In black soils, the infiltration of water is slow and much of the rainfall is lost as runoff. To improve infiltration and storage of rain water in these soils, vertical mulches are formed. It consists of digging narrow trenches across the slope at intervals and placing the straw or crop residues in these trenches. This is mostly practiced in coffee gardens. The pruned plant material is placed in contour trenches formed between rows of coffee or in trenches around the plants in concentric circle each year in one circle.

(vi) Live Mulching: It is the term used to describe the covering of soil surface through the plant canopy in intercropping system e.g. sorghum + fodder cowpea.

(vii) Pebble Mulch: Where small pebbles like stone are placed on the soil surface. This mulching will be successful in dryland fruit tree culture. The pebbles placed on the basin of trees not only reduce evaporation but also facilitate infiltration of rain water into the basin.

4.2.2 Effect of Mulches on Soil Properties

(i) **Soil Structure:** Surface mulches reduce the impact of falling rain drops, thus reducing dispersion and sealing of the soil pores leading to crust formation. Hence, the soil structure is protected. The mulches also improve soil structure due to decomposition of mulch.

(ii) **Soil Salinity:** Under dry land conditions due to limited precipitation, soluble salts move only to a limited depth and readily return to the surface as the soil water evaporates. Due to salt accumulation in surface layers the germination and seedling establishment may be adversely affected. Hence mulches will reduce soil salinity problem by increasing infiltration and reducing evaporation.

(iii) **Soil Water:** The soil moisture content is improved by induced infiltration, reduced evaporation and reduced transpiration by weeds. Surface mulches also obstruct the free exchange of water vapour from soil surface into the atmosphere and hence increase soil water content.

(iv) **Soil Temperature:** The effects of mulches on soil temperature are highly variable and depend on the type of mulch material. White or reflective type of plastic mulches generally decreases soil temperature, while black plastic mulches may increase soil temperature. Crop residues moderate temperature by decreasing it in summer and by increasing in winter season. This is due to combined effect of radiation interception and evaporative cooling.

(v) **Soil Erosion:** The ease by which soil particles are moved by wind and water is related to size of soil particles and wind and water velocity. The particles of size greater than 0. 84 mm in diameter are generally not eroded by wind but they are easily eroded by water. The mulches reduce the direct impact of falling raindrops on soil thereby preventing soil dispersion and consequent sealing of soil pores leading to soil erosion.

4.3 Use of Anti–transpirants

Nearly 99 per cent of water absorbed by the plants is lost in transpiration. If transpiration is controlled, it may help in maintance of favourable water balance in the plant system. Anti-transpirant is any materiel applied to transpiring plant surfaces for reducing water loss from the plant. The four types of antitranspirants used in crop production are as follows:

(i) **Stomatal Closing Type:** Most of the transpiration occurs through the stomata on the leaf surface. Some fungicides like phenyl mercuric acetate (PMA) and herbicides like atrazine in low concentrations serve as anti–transpirants by inducing stomatal closing. These might also reduce the photosynthesis simultaneously. PMA was found to decrease transpiration to greater degree than photosynthesis in a number of plants. Though, success was reported from glass house studies, their effectiveness under field conditions is limited.

(ii) **Film Forming Type:** Plastic and waxy materials which form a thin film on the leaf surface retard the escape of water due to formation of physical barrier. Mobileaf, hexadecanol, silicon are some of the film forming type of anti-transpirants. The success of these chemicals is limited since they also reduce photosynthesis. The desirable characteristics of film forming type of anti-transpirants are: they should form a thin layer, they should be more resistant to the passage of water vapour than carbon dioxide and the film should maintain continuity and should not break.

(iii) **Reflectant Type:** These are white materials which form a coating on the leaves and increase the leaf reflectance (albedo). By reflecting the radiation, they reduce leaf temperatures and vapour pressure gradient from leaf to atmosphere and thus reduce transpiration. Application of 5 per cent kaolin spray reduces transpiration losses. A diatomaceous earth product (celite) also increases reflection of solar radiation from crop canopy.

(iv) **Growth Retardants:** These chemicals reduce shoot growth and increase root growth and thus enable the plants to resist drought. They may also induce stomatal closure. Cycocel is one such chemical useful for improving water status in the plant.

Anti-transpirants generally reduce photosynthesis. Therefore, their use is limited to save the crop from death under severe moisture stress. If crop revives, it can utilise the rainfall that is received subsequently. Anti-transpirants are also useful for reducing the transplantation shock of nursery plants. They have some practical use in nurseries and horticultural crops.

4.4 Other Practices

4.4.1 Wind Breaks and Shelterbelts

Wind breaks and shelterbelts are permanent vegetative measures for controlling evaporation from water and soil surfaces. The wind break is any type of barrier either mechanical or vegetative like buildings, orchards or farmsteads. A shelterbelt is the longer barrier than wind break and consists of shrubs and trees. They exert a wider influence than wind breaks. The width of shelterbelt may vary from one to five rows or even more. These are effective for increasing the air resistance to water vapour transfer. Wind strip cropping consisting of growing annuals between strips of perennial grasses has also been found effective in conserving moisture at CAZRI, Jodhpur.

The direction from which wind is blowing is called windward side and direction to which wind is blowing is called leeward side. Shelterbelts are planted across the direction of wind. They do not obstruct the wind flow completely. Depending upon their porosity, certain amount of wind passes through the shelterbelts while the rest defects and crosses over the shelterbelts. It thus reduces wind speed without causing turbulence. The protection offered by the shelterbelts is dependent on the height of central tree row in the shelterbelts. Generally, shelterbelts give protection from desiccating winds to the extent of 5 to 10 times of their height on windward side and up to 30 times on leeward side. Due to reduction in wind speed, evaporation losses are reduced and more water is available for plants. The beneficial effect of shelterbelts is seen clearly in drought years. In addition, shelterbelts reduce erosion.

4.4.2 Alley Cropping

In this arable food crops are grown in the alleys formed by hedgerows of perennial trees or shrubs. A similar effect on reduction in atmospheric water demand and transpiration as described under shelterbelts is caused in alley cropping. This method is also called as hedge row intercropping. The best plants for this are subabul and glyricidia because they grow well in degraded soils. e.g. in the hedge rows of subabul, blackgram, cowpea, sunflower *or* groundnut can be intercropped.

4.4.3 Weed Control

Weeds have a high transpiration coefficient than most of crops and transpire more water to produce a unit quantity of dry matter compared to crops. Early weed control prevents unwanted transpiration loss through weeds.

♦

5

WATER HARVESTING

In dryland ecosystem the soil topographic features and climatic factors prevailing are highly conducive to generate runoff. Further intense and heavy showers are the characteristics of tropical areas. The combination of these factors results in excessive runoff from the fields with the soil profile generally remaining unsaturated. Research conducted in several areas of drylands in the country and elsewhere has revealed the feasibility of water harvesting and storing excess runoff in small dugout ponds and reutilizing the same for providing supplemental/ life saving irrigation to crops during prolonged periods of drought for enhancing productivity of crops.

Water harvesting is the process of collecting, conveying and storing of rainwater (this is the most practical and feasible in Indian conditions) that runoff from natural or man made catchment surface. So what we need to practice in India is rain water harvesting, though it would suffice to call it water harvesting. The problem of large variation in water availability and growing demand leads to the need for water harvesting. The potential of water harvesting can be illustrated by pointing out that one millimeter of rain equals10,000 litre of water per hectare. It becomes apparent that a small area of impermeable surface can collect a relatively large volume of water. Water harvesting will not be feasible in some areas. On the other hand, there are many areas where water harvesting offers the only reasonable opportunity to develop new water supplies.

5.1 Definition

Water harvesting is defined as **"collection of runoff for its productive use"**. Water harvesting is hydro agronomic term covering a whole range of methods of collecting and concentrating various forms of runoff. In the water harvesting process, the runoff producing area and part of the land and most of the rain water will become productive. More importantly, agricultural production becomes possible and water harvesting systems might be for a single purpose or for multipurpose uses. Water harvesting is also defined as a water management technique for growing crops in arid and semi-arid areas where rainfall is inadequate for dryland crop production and irrigation water is unavailable.

Water harvesting is an broader term covering a range of methods of collecting and conserving various forms of runoff from different sources. Water harvesting can be done both *in-situ* or *ex-situ* (reservoir or dugout ponds) depending on the situation as well as socio-economic conditions of the farmer. In dryland agriculture, water harvesting usually, denotes the collection of excess runoff in a storage tank and using for protective/ life saving irrigation at times of prolonged drought during crop season. In this chapter only *ex-situ* water harvesting techniques shall be covered as in-situ moisture conservation practices are dealt in previous chapter.

5.2 Components of Water Harvesting System

Principle components of water harvesting for crop production include all methods of collecting runoff for productive purposes as illustrated in fig. 5.1. All water harvesting systems have three components given as under-

(i) **The Catchment Area/ Runoff Area** - The catchment area is the part of land that contributes the rain water. It varies from a few square meters (micro catchment) to as large as several square kilo meters (macro catchment).

(ii) **The Storage Facility** - The storage facilty is a place where the runoff water is stored from the time it is collected until it is utilized by field crops, animals, human beings and/ or other purposes.

Storage can be:

- Above the soil surface as in surface reservoirs or ponds.
- In the soil profile as soil moisture and / or
- Underground in cisterns or as ground water in aquifers.

(iii) **The Command Area** - The command area is where water is used.

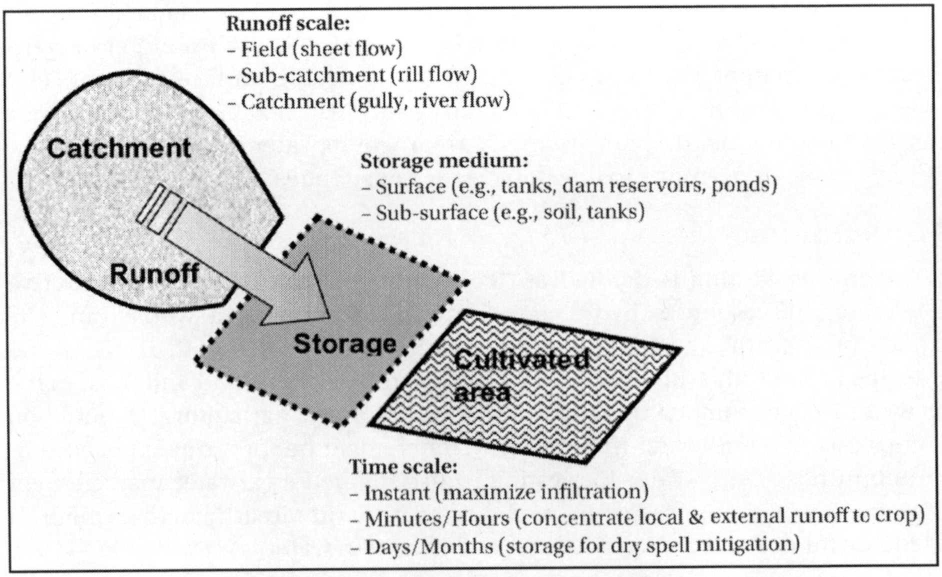

Fig. 5.1: Water Harvesting System

5.3 Water Harvesting Techniques

Water harvesting techniques may be classified into two categories *i.e.* micro catchment and macro catchment as follows–

(i) Micro Catchment Techniques: These are the techniques that directly supply runoff water from a small catchment to the crop and thus water accumulates around the plant, infiltrates into the soil and is stored in the crop root zone. These are called micro catchment techniques, because the runoff yielding catchments are usually small and directly adjacent to the targeted crop. Widely used micro catchment techniques like contour bunding, stone wall terracing, bench terracing, contour border strips, tillage practices, compartmental bunding, furrowing, broad bed furrow and ridges, etc., These techniques are discussed in the previous chapter.

(ii) Macro Catchment Techniques: These techniques concentrate rainwater runoff flowing in an ephemeral wadi (natural channel) and store it in a prepared storage facility such as a reservoir/farm pond for subsequent beneficial use. This category also includes macro catchment in which water is diverted by proper damming or cross structure out of the wadi course to inundate nearby lands.

5.4 Methods of Water Harvesting

5.4.1 Traditional Methods

Water harvesting is an age old practice in India. A number of water harvesting techniques prevailed in different parts of the country, which were well adopted in the local environment. Some of the water harvesting systems, common in India are presented as under–

Artificial glaciers – These are made close to the village, to ensure water before the natural glaciers melt in late summer.

Beris – Collects water within the desert sand.

Eris – Water catches (tanks) built into the terrain collecting surface runoff. Overflow from eris at higher levels travels into the lower ones.

Johads – Reservoirs behind small dams constructed across the slope to collect runoff and force it to infiltrate.

Khadins – It is a unique land use system wherein runoff water from rocky catchments is collected in valley basins during rainy season. Crops are grown in winter season when water is receded in shallow pond on the residual moisture.

Kui – Collects rain water trapped in the sand above a hard gypsum layer. They are usually covered with planks of wood.

Kuls – These are diversion channels for irrigation. They carry water from glaciers to villages.

Kundis – Covered dome shaped storage tanks holding rainwater for the local people and animals in Rajasthan.

Mughal – Ground water based water supply system in Madhya Pradesh.

Virdas – Shallow wells dug in low depressions to collect rainwater, used by the nomadic Maldharis.

Zabo system – Below a catchment area, kept under natural vegetation, are ponds dug to harvest water for irrigation and animal consumption. It is common to maintain a fenced cattle yard below the ponds and above the paddy fields. Thus, runoff water enters the fields rich in manure.

Nadi or village pond – It is constructed for storing water from natural catchments. The capacity of *nadis* ranges from 1200 m^3 to 15000 m^3.

Tanka – It is an underground tank or cistern constructed for collection and storage of runoff water from natural catchment or artificially prepared catchment or from a roof top. The vertical walls are lined with stone masonry or cement concrete and the base with 10 cm thick concrete. The capacity of tank ranges from 1000–6,00,000 litre.

Pynes – These are channels constructed to utilize water flowing through hilly rivers.

5.4.2 Modern Methods

5.4.2.1 Farm Ponds

Farm ponds are small tanks or reservoirs constructed for the purpose of storing water essentially from surface runoff. These ponds are useful for providing supplemental life saving irrigation to the crops, drinking purposes, supply of water to livestock, fish rearing, etc. A large number of ponds constructed in the catchment area reduce the flood flows downstream. These are also used for storing rainwater, which is subsequently used for supplemental or life saving irrigation particularly in the condition of long dry spells. Thus a farm pond has a significant role in the area of rainfed/dryland agriculture. The design and construction of farm ponds require a thorough knowledge of the site conditions and water requirements. Some sites are ideally suited for locating the ponds and for taking the advantage of natural conditions, should always be considered for economic consideration.

5.4.2.2 Types of Farm Ponds

(a) Embankment Type

These ponds are generally constructed across the stream or water course. Such ponds consist of an earthen dam, whose dimensions are estimated on the basis of volume of water to be stored, etc. These farm ponds are usually built in the area where land slope ranges from gentle to moderately steep; and also where stream valleys are having sufficient depths to permit a maximum storage volume with least earthwork required.

(i) Percolation pond

(ii) Dry stone masonary pond

(iii) Cement machinery pond

(iv) Anicut

(v) Check dam

(i) **Percolation Pond:** These are constructed across natural streams and nalla to collect and impound the surface runoff from the catchments during monsoon rains and store it for longer time to facilitate infiltration and percolation of water into the sub-strata to raise the ground water level in the zone of influence of pond (Fig. 5.2). They should fulfill following technical considerations:

- No percolation pond should be located in heavy soils or in impervious strata and be proposed at places where top soil is highly porous.
- The size of the pond should be decided on the basis of expected runoff from dependable monsoon rainfall in the catchment area.

Fig. 5.2: Percolation Pond

- There should be adequate cultivable land and enough number of irrigation wells or there should be scope for putting up irrigation wells in the zone of influence of pond.
- The topography of the proposed site of the pond should be such that naturally a deep pond is formed so that evaporation losses could be reduced.
- Suitable and adequate soil for forming embankment should be available within the pond area and should not be a need for transport of soil from outside.

(ii) **Dry Stone Masonry Pond:** Dry stone masonry ponds of 1, or 2, or 2.5 m height are constructed by the farmers to harvest rainwater. In this type of pond the upstream and downstream walls are made up of dry stone and murrum/soil is filled in between them at an appropriate interval to provide stability to the structure. The design dimensions of upstream wall and downstream wall are taken same. Dry stone masonry ponds should not be constructed for the height more than 2.5 m because for the height more than this, the structure will overturn due to water pressure. A provision of waste weir for safe disposal of excess water is necessary for such ponds. It does not require any technical skill and can be constructed by the famer in a narrow valley having the catchment area maximum upto 50 ha (Fig. 5.3).

Fig. 5.3: Dry Stone Masonry Pond

(iii) **Cement Masonry Pond:** Cement masonry pond, a low cost water harvesting structure upto the height of 3 m is indigenously constructed by the farmers. In this type of pond the upstream wall and downstream wall is constructed with cement masonry. In between two walls murrum soil is filled at an appropriate interval to provide stability to the structure. The design dimensions of upstream and downstream walls are taken according to the height of the structure but in any case the bottom width should not be more than 2.0 m. At the top of the head wall plain cement concrete is provided to join both the walls.

It is dug cum embankment type pond constructed in the narrow valley portion of stream of nalla particularly at the site, which already has depressions. These ponds are being widely constructed under NWDPRA for storing rainwater for the purpose of drinking water for livestock and ground water recharge.

(iv) **Anicut:** Anicut is weir structure constructed across the natural drain or nalla for impounding the water. It is a masonry structure, which intercepts the water from local catchments and store it for optimum utilization. The word anicut is a local name of this water harvesting structure commonly known in Rajasthan. The water store behind the structure can be used for supplemental irrigation to the crops and also provide drinking water for human and livestock. This will also help in recharging of the wells located downstream side of the structure. Selection of suitable site of "Anicut" is very important as the cost of construction as well as storage volume depend upon the site (Fig. 5.4).

Fig. 5.4: Anicut

(v) **Check Dam:** A check dam is a small, sometimes temporary dam constructed across a swale, drainage ditch, shallow rivers or water way to counter act erosion by reducing water flow velocity. The small dams retain excess water flow during monsoon rains in a small catchment area behind the structure. Pressure created in the catchment area help to force the impounded water into the growled (Fig. 5.5). The basic purpose of such dams is to recharge replenishment of ground water reserve and wells. The water stored in the dam is primarily used for irrigation purposes, but can also be used for livestock and domestic needs.

Fig. 5.5: Check Dam

(b) Excavated or Dugout Type

These ponds are constructed by excavating the soil to its desired depth /capacity. The use of this type of pond is suitable particularly where a small supply of water is required. Such ponds are also divided into two types viz., spring creek fed and off-stream pond depending upon the sources of water available for feeding them. Dug out or excavated ponds are constructed in areas of flat or gentally sloping land not suited for ponds with dams. As the name implies, dug-out ponds are created by removing soil and allowing water to fill in the dugout area (Fig. 5.6). In case of dugout ponds, if the stored water is used for irrigation, the water has to be pumped out.

Fig 5.6: Dugout Pond

(i) **Spring or Creek Fed Ponds:** These ponds are those where a spring or creek is the source of water supply to the pond. Such ponds are generally found in hilly areas where natural spring or creeks are available.

(ii) **Off-stream Storage Ponds:** This type of ponds are constructed by the sides of ephemeral stream, which flows seasonally. The idea is to store the water obtained from the seasonal flow in the streams. Suitable arrangement needs to be made for conveying the water from the stream to the storage ponds.

To design a dugout pond the rainfall, runoff, evaporation and seepage data are essentially required, but these are mostly not available for most of the farm sites. If these are available, then cost of an earth shine hydrologic investigation is not required, however, based on few main factors affecting the pond design, a general guide line has been developed.

5.5 Selection of Site for Construction of A Farm Pond

(i) It should not be located on very sandy or shallow soils.

(ii) It should be in accordance with the topography of the area to get required water storage.

(iii) It should have a large contributing surface runoff land.

(iv) It should preferably be a permanent grass or tree areas rather than cultivable lands and lands prone to soil erosion.

5.6 Size of Farm Pond

Farm ponds having a capacity of 250 to 300 m^3 for each hectare of land is ideal to collect the runoff. For a four hectare farm, the size of farm pond may be 20 m × 20 m with a depth of 3 m in clay soil. Under red soils, the pond may be lined with soil and cement in the ratio of 8:1 and thickness of 5 cm for the purpose of preventing initial seepage losses.

5.7 Seepage Control in Pond

Seepage is counted as one of the important factors for pond design. In the areas of high gypsum strata the seepage loss is much more, causing construction of farm pond impractical. Although, everywhere there is seepage due to ponding of water but it should be within limit. Normally seepage loss at the rate of one meter depth of water per year is allowed. Various seepage control methods have been tested in the field and research projects. Interestingly, more effectiveness involves more cost, which may not prove economical under any circumstances. However, depending upon farming activity and economic condition of farmer, the following methods may be tried:

(i) **Puddling:** Seepage rate of nearly constructed ponds or tanks can be reduced considerably by puddling the floor of pond by man or animal or both.

(ii) **Compaction:** In well graded soil, compaction of bed of the structure results in sizable reduction of seepage rates. The top 20 to 25 cm depth of soil is scarified, pulverized and then compacted at optimum moisture content. The thickness of compacted seal depends upon the head of water to be stored. But a 15 cm thick layer is generally maintained.

(iii) **Clay Layer:** A clay layer is used to cover the porous bottom of a pond or tank or other such structures to make them leak proof. For three meter depth of stored water a 25 cm thick clay blanket may be useful.

(iv) **Bentonite Treatment:** Bentonite is a very fine textured colloidal clay, which is mixed with surface soil of the pond and compacted.

(v) **Addition of Sodium Salt:** Addition of deflocculated chemical like sodium carbonate, sodium chloride and tetra sodium polyphosphate to the surface soil of the reservoirs bed/ deflocculated soil of the bed. Thus the seepage rates can be considerably reduced. For these substances to be effective the clay content of the soil should be at least 15 per cent and the cation exchange capacity should be greater than 15 meq per 100 g. Sodium carbonate which is more durable, generally applied at the rate of 0.48 to 1.45 kgm^2. After every 2 to 3 yr an extra amount of sodium carbonate is to be added to continue the effectiveness.

(vi) **Soil-cement Lining:** Well graded sandy soils mixed with cement in the ratio of about 10–12: 1 by volume are found to reduce the seepage rates from ponds or tanks considerably. The higher is the cement content the greater will be the effectiveness. It requires proper watering and curing for good setting and longer life.

(vii) **Asphalt Lining:** This lining consist of the mixture of sand and gravel together with asphalted cement. This water tight layer is not costly but encourages growth of vegetation which easily penetrate the lining making them defunct rapidly.

(viii) **Stone Slab and Brick Lining:** Where stones are available abundantly they are laid in cement mortar (1:3) on the bed and sides of the pond, which are quite effective in reducing seepage loss. Bricks can be used instead of stone but a lining with rich cement mortar (1 cm thick) on the surface of that lining is required to obtain desirable effectiveness.

(ix) **Cement Concrete Lining:** It is the most effective method of seepage control, but the initial cost is very high and therefore it is uneconomical for common farming purpose. A cement concrete mixture is applied at thickness of 3 to 4 cm. Measures should be taken against the cracking of this layer through proper curing and maintenance in due course.

(x) **Polythene, Plastic, Rubber Films:** The films or sheet of these materials can be used as open or buried layer in the bottom and side surfaces of reservoirs and waterways to cut off the seepage through them. It shows varying degree of success, however, they require special precautions against punching or cutting, rodent burrowing and weathering.

5.8 Precautions in Construction of Farm Ponds

(i) Downstream slope of the pond be protected with gravel or vegetative barrier to avoid erosion.

(ii) A spill way must be provided for emergency purposes.

(iii) A permanent fence (if possible) may be built around the pond to prevent damage.

Since the stored water is limited, steps should be taken to reduce losses from seepage and evaporation.

5.9 Water Harvesting and Life Saving Irrigation

Rainwater is the key input in dryland agriculture. In a tropical country such as India, which experiences extreme variation in rainfall in both space and time, rainwater management assumes vital importance in cutting down risks and stabilizing crop production in dry areas. When rains are received with an intensity far reaching infiltration rate, runoff is inevitable. It varies from 10 to 40 per cent of total rainfall. Of this at least 30 per cent can be harvested into water storage structures.

5.9.1 Supplemental Irrigation / Life Saving Irrigation

The runoff collected from different water storage structures is of immense use for protecting the dryland crops from soil moisture stress during prolonged dry spells. Supplemental or life saving or protective irrigation is given to sustain the dry land crops during the drought periods and take the advantage of subsequent rains. In dry areas, water, not land is the most limiting resource for crop production. Maximizing the water productivity but not the yield per unit land is the better strategy for dry farming areas. Supplemental irrigation is a highly efficient practice for increasing productivity of crops in arid regions. The response to supplemental irrigation varies with crops, time of irrigation, depth of irrigation, method of water application and fertilizer application.

5.9.2 Quantity of Irrigation Water

Crops differ in responding to amount of irrigation water by supplemental irrigation during the dry spell. Groundnut responds to 10 mm of irrigation through sprinkler on alfisols during pod development stage. Cotton needs a minimum of 30 mm of water to respond to irrigation applied either by sprinkler or drip irrigation system on vertisols. Chickpea similarly need 30 to 40 mm of supplemental irrigation applied as drip or sprinkler irrigation during flowering. Pigeonpea responds to 20 mm irrigation water applied at pod development stage with drip irrigation. Irrigation can be provided near the row, covering about 20 per cent of the cropped area, leaving 80 per cent of inter row zone. Pot watering, applying small quantity of water (around 250 ml) manually to each hill, is highly useful either for sowing or for transplanting in widely spaced crops like cotton, pigeonpea, tomato, tobacco etc. Similarly, pot watering to protect the seedlings during early crop growth stage is highly useful. The amount of water, if calculated over the entire area, works out less than 0.5 mm/ha. For example, pot watering of cotton seedlings at 250 ml per hill works out to 5,000 litre per ha which works out to 0.5 mm per ha. Productivity of harvested water can be increased by applying small quantity of water to large areas than heavy irrigation to small area. If rains occur immediately after irrigation, there will be no impact of irrigation and in black soil, it may reduce the yield.

5.9.3 Time of Irrigation

Unlike in irrigated agriculture, the critical stage concept does not suit well, as dry spell may reduce the growth and yield of crop at any stage. Vegetative stage is considered as non- critical stage in irrigated agriculture but in arid regions, dry spell during vegetative stage prolongs the crop duration, which may ultimately result in crop failure due to late season drought. Death of seedlings also cause reduction in yield due to dry spell in vegetative stage, therefore, the strategy for getting successful crop is providing small quality of water, if available, at any stage if the dry spell is more than 10 days in light soils and 15 days in heavy soils.

5.9.4 Method of Irrigation

Surface methods of irrigation like check basin and furrow methods are not suitable for supplemental irrigation, mainly for three reasons: the rainfed lands are uneven, conveyance losses may go up to 30 per cent and limited amount of water available for irrigation. Drip and sprinkler irrigations are more suitable because small amount of water can be delivered, even on uneven soils without conveyance losses. Subsurface drip irrigation is very efficient for providing supplemental irrigation. The main drawback of micro-irrigation system is high initial cost of the system. Pot watering is another efficient method being used by the farmers for transplanting crops like tobacco, chilli, tomato, etc.

Dry land farming, particularly cultivation of crops, frequently sufters from shortage of soil moisture in critical periods of crop growth. An arrangement of storing of protective or supplemental irrigation water in different water harvesting structure particularly in farm ponds and application of the same at the critical juncture helps to boost the yields of cereal crops up to 40 to 60 per cent. However, due to high cost of the supplemental irrigation and poor farm-gate price of field crops frequently established it as an uneconomic venture. Vegetables like chilli, brinjal, tomato, bottle gourd, etc. also yield more with supplemental irrigation and economically better option than cereals. The protective or supplemental irrigation to different perennial fruit crops suitable in dry areas, like pomegranate and ber proved highly rewarding.

◆

6

CROP MANAGEMENT PRACTICES FOR RAINFED AGRICULTURE

In India about 60 per cent of total net sown area comes under drylands, contributing nearly 44 per cent food production and supports about 40 per cent of the human and 60 per cent of livestock population. As regards the crops, 80 per cent of maize, 95 per cent pearlmillet and sorghum, 91per cent of pulses, 80 per cent of oilseeds, nearly 65 per cent of cotton and almost whole of jute and mestha, about 50 per cent upland rice and 19 per cent of wheat are grown under rainfed conditions. Productivity of crops in dryland farming is low due to inadequate supply of moisture and other constraints. Since production gains from irrigated lands cannot meet the growing demands for ever increasing population of the country, more emphasis needs to be given for dry land agriculture. To increase the yield of different crops grown under rainfed/dryland conditions, following improved agro-technologies needs to be adopted.

6.1 Selection of Suitable Crops and Cropping Systems

The probability of occurrence of different amounts of rainfall and the corresponding length of growing season could be taken as criteria for selecting suitable crops and cropping systems in any season. Generally adaptable crops for dry regions are cereals and millets such as rice, wheat, maize, barley, oats, sorghum, pearlmillet and fingermillet, oilseeds such as groundnut, sesame, castor, safflower, sunflower, linseed, rapeseed & mustard, niger and taramira, pulses such as pigeonpea, blackgram, greengram, chickpea, cowpea, and other crops such as clusterbean, colocasia, tobacco, and vegetables. Among fodder crops non-legume fodder crops such as sorghum, pearlmillet, maize, sanwa (*Echinochloa* spp.), dinanath grass and legume fodders such as cowpea, clusterbean, fieldbean and ricebean are important. Pasture legumes such as siratro and stylo are also used as green fodder or as hay.

Cropping systems differ according to climate and soil types. In regions of low rainfall (400 mm), a short growing season (60–75 d) with shallow depth of soil as in western Rajasthan and parts of Haryana and Kutch, grass is the principal crop. In addition to grasses, short duration pulses or agroforesty based systems might hold some promise in such areas. With increasing rainfall and a longer growing season, cereals such as pearlmillet, sorghum, maize or upland rice may be grown. In areas with 400 to 750 mm annual rainfall, monocropping with traditional long duration crops is common. Even in areas with an annual rainfall above 750 mm, soil storage capacity of 150 to 200 mm of available moisture, farmers have been growing either a rainy season crop on residual soil moisture after fallowing during the rainy season. Similarly when the rainfall is between 500 to 750 mm with a distinct period of moisture surplus, the intercropping facilitates the growing of either cereal + legume or legume + legume in a system of different maturity patterns. With the recent emphasis on intercropping system, it has now become possible to harvest almost the full yield of a cereal component and extra yield to the tune of 60 per cent of the associate legume crop. Some examples of intercropping systems under drylands are given as under in table 6.1.

Table 6.1: Intercropping for Different Regions in India

Region	Intercropping system
Northern Region	
Samba (Jammu)	Castor + blackgram/ sesame/ cowpea/ sunflower
	Maize + cowpea (fodder)
	Maize + greengram
Hissar (Haryana)	Pearlmillet + clusterbean / greengram
Central Region	
Dehradun (Uttarakhand)	Pigeonpea + soybean
	Maize + pigeonpea
Agra (U.P.)	Pearlmillet + blackgram/ greengram
	Pigeonpea + greengram/ pearlmillet
Varanasi (U.P.)	Pigeonpea + maize/ blackgram/ millet
Eastern Region	
Ranchi (Bihar)	Barley + rapeseed
	Wheat + bengalgram/ rapeseed/ lentil
Bhubaneswar (Orissa)	Pigeonpea + blackgram/ sorghum/ groundnut
Rewa region (M.P.)	Pigeonpea + sorghum + fingermillet/ blackgram/ sunflower
Jhansi (M.P)	Sorghum + blackgram/ greengram / soybean
	Wheat + bengalgram/ mustard
	Barley + bengalgram/ mustard
	Bengalgram + linseed/ mustard
Indore (M.P)	Pigeonpea + sorghum, cotton + soybean

Region	Intercropping system
Western Region	
Udaipur (Rajasthan)	Sorghum + greengram
	Maize + blackgram
	Maize + pigeonpea
	Groundnut + sesame
	Castor + greengram
	Chickpea + mustard
	Chickpea + safflower
	Pearlmillet + blackgram
Anand (Gujarat)	Pearlmillet + pigeonpea / castor/ sunflower
Akola (Maharashtra)	Pigeonpea + blackgram / greengram / sorghum/ groundnut, cotton +sunflower/ blackgram / groundnut
Sholapur (Maharashtra)	Pigeonpea + pearlmillet / sorghum/ blackgram
Southern Region	
Anantapur (A.P.)	Cotton + greengram / pearlmillet/ sorghum / foxtailmillet / groundnut/ sunflower
Bijapur (Karnataka)	Cotton + groundnut / sorghum/ pearlmillet / greengram
Mysore region (Karnataka)	Pigeonpea + pearlmillet
	Cotton + pearlmillet
	Groundnut + cotton
Kovilipatti. (T.N.)	Cotton + blackgram / greengram
	Sorghum + blackgram/ cowpea / *Dolichos lablab*
	Pearlmillet + greengram/ clusterbean / cowpea /lablab bean

In areas with more than 750 mm annual rainfall with a soil storage capacity of 150 mm or more of available moisture, sequential cropping is possible. In the Jodhpur belt of Rajasthan pearlmillet followed by chickpea and greengram followed by mustard, in the vertisols of Hyderabad - sorghum - safflower, sorghum – chickpea and maize– chickpea, in Bangalore–cowpea– fingermillet, in Solapur–pearlmillet–chickpea and greengram *–rabi* sorghum, in Bijapur–greengram followed by safflower are dominant cropping sequences. Cropping system suitable for different regions of the country are given in table 6.2 as under.

Table 6.2: Cropping Systems for Different Regions of India under Dryland Agriculture

Region/soil type	Cropping system	
	Kharif	*Rabi*
Northern Region		
Samba (Jammu)	Maize Greengram Cowpea Sunflower Groundnut Pearlmillet	Wheat Barley Wheat Wheat/ barley Barley Bengalgram
Punjab	Sorghum Maize	Wheat Wheat/potato/ barley
Hissar (Haryana) Arid Soils (Sierozemic Soils)	Pearlmillet Clusterbean	Clusterbean / chickpea Chickpea
Central Region		
Dehradun (Uttarakhand)	Maize Upland Rice	Wheat Wheat/ bengalgram
Agra (U.P.) (Entisols)	Pearlmillet	Mustard Chickpea / barely
Varanasi (U.P.) (Entisols alluvial soils)	Sesame Rice Pearlmillet	Chickpea / mustard Chickpea / mustard Chickpea /mustard
Eastern Region		
Ranchi (Bihar) (Alfisols and related red soils)	Rice/ maize	Chickpea/ linseed / barley /rapeseed
Bhubaneswar (Orissa) (Alfisols and related red soils)	Rice /finger millet/maize	Linseed/ mustard
Rewa region (M.P.) (Vertisols and related black soils)	Rice Sorghum Greengram	Chickpea/ lentil Wheat Chickpea
Jhansi region (M.P.) (Vertisols and related black soils)	Sorghum/Pearlmillet/ Blackgram/Sesame	Safflower/ chickpea
Indore (M.P.) (Vertisols and related black soil)	Maize Soybean Soybean	Chickpea Safflower Wheat
Western Region		
Udaipur (Rajasthan) (Vertisols and related black soils)	Sorghum Maize Maize Blackgram	Safflower Chickpea Mustard / taramira Chickpea /wheat
Anand (Gujarat)	Pearlmillet / pigeonpea castor/ sunflower	Wheat

Region/soil type	Cropping system	
	Kharif	***Rabi***
Akola (Maharashtra)	Greengram/ sorghum	Safflower
Sholapur (Maharashtra)	Greengram Pearlmillet Blackgram	Safflower Chickpea Sorghum
Southern Region		
Anantapur (AP)	Sorghum Pearlmillet Blackgram Greengram Cowpea	Safflower/ horsegram Cowpea/ blackgram Sorghum Pearlmillet Pearlmillet / sorghum
	Mono cropping is generally adopted	
Hyderabad (A.P.) (Alfisols)	Pearlmillet Sorghum	Cowpea Horsegram
Bijapur (Vertisols)	Sorghum Greengram	Greengram /blackgram Sorghum / safflower
Bangalore (Karnataka) (Alfisols)	Cowpea Cowpea	Fingermillet Chilli
Mysore (Karnataka)	Groundnut Setaria Hybrid pearlmillet	Sorghum Safflower Cotton
Bellary (Karnataka)	Cotton Blackgram	Sorghum Bengalgram
Kovilipatti (TN)	Cotton/Sorghum/ Chilli	Pearlmillet/cowpea/ horsegram

6.2 Selection of Suitable Varieties

In addition to suitable crops and cropping systems, adoption of suitable crop varieties are essential prerequisite for making the dryland farming a successful enterprise. According to breeders all ideal ideotypes for dryland agriculture should have such characters that they can evade drought like – early in growth duration and early vigour, deeper root system with maximum ramification of roots at deeper zone, thick stems, moderate tillering with synchronous growth habit, resistance to diseases, lesser period between flowering and maturity and effective photosynthetic behavior with greater sink capacity. Some of the crop varieties recommended in different dryland areas of the country are given as under in table 6.3.

Table: 6.3: Important Varieties of Different Crops Suitable for Dryland Agriculture

Crop	Suitable varieties
Rice	Pusa 2 – 21 , Bala
Maize	Navjot, GM - 6, Pratap Makka - 3, Pratap Makka - 5, PEHM - 3, HIM - 129

Crop	Suitable varieties
Sorghum	CSH - 14 , CSH - 15, Pratap Jowar - 1430, SPV - 837, CSV -17, SSG - 59 - 3 (fodder)
Pearlmillet	HHB - 67, HHB - 146, Pusa Composite - 334, GHB - 757, GHB -558
Clusterbean	RGC - 1031, RGC - 1017, RGC - 1055, RGC - 936
Pigeonpea	ICPL - 151 , ICPL - 87, Gujarat Tur - 100
Blackgram	Pant U - 19, KU - 96 - 3, TAU - 2 , T - 9
Greengram	RMG - 62, K - 851, Pant Mung - 4
Cowpea	RC - 101, CO - 6, C - 152, Gujarat Cowpea - 4
Kidneybean	PCMO - 880, PCMO - 186, PCMO - 34
Groundnut	JL - 24, GG - 2 , TAG - 24, TG - 37A, JGN -3, K -134, Kadiri -6
Sesamum	RT - 46, RT -125, Gujarat Til - 2
Castor	Aruna , Gauch - 1 , Gauch - 4
Soybean	NRC -7, JS 71-05
Wheat	GW - 273, GW - 1139, Lok -1, RAJ - 3765, HI - 1500, HI-1531, Raj. - 3077
Barley	RD - 2508, RD - 2503, RD - 2052
Chickpea	Pratap Raj Channa, ICCV-10 , RMG - 888, RSG -963
Mustard	Vasundhara, Pusa Jaikishan, RH - 30, Pusa Bahar, Shivani
Taramira	RTM - 314 , T 27

6.3 Sowing Time

The ideal time for sowing in dryland agriculture is found to be early part of the sowing season. It appears most crucial and loss in productivity as a result of delayed sowing and which cannot be compensated by any other improved agronomic practice. Early sowing has several advantages such as improved water use efficiency, good early seedling vigour and longer growing season, provide opportune time for second crop in double cropping systems and avoidance of cooler weather at flowering for *kharif* crops in north India. Hence, success of crops in dry land largely depends on time of seeding especially with improved cultivation. Delayed sowing leads to reduced length of growing period and exposure of crop to terminal stress. The very first showers, therefore, should be utilized for sowing by keeping the field ready prior to probable rainfall date.

Prior to sowing it is necessary to ensure that 20 cm soil is wet and on red soils seeding operation should be carried only upto two days after the monsoon showers. Weather forecast should be followed to avoid risk.

Sowing time alone can make a lot of difference in production and productivity of dry land crops. Sowing time is not only crucial from moisture harvesting, crop stand establishment and land productivity point of views but it has long range effect on total productivity also. An increase in yield has been found by advancing the date of sowing of *kharif* and *rabi* crops. Besides better yield, an early sowing of *kharif* crops help in proper establishment of plants to safeguard against torrential rains or temporary water logging. Early sowing of *kharif* crops results in their early maturity and thereby it facilitates an early planting of *rabi* crops. Early planting of *rabi* crop, after an early harvest of *kharif* crop, helps in overcoming the moisture stress at later stages of plant growth, particularly at grain filling stage.

6.4 Seeding Techniques

Seeding operation is an important practice which helps in obtaining good stand establishment, early vigour and better yield of crops. Line sowing, dry seeding, depth of seeding and seed hardening are of considerable importance are described as under:

(i) **Line Sowing:** Seed drill sowing is more useful in view of precision in the depth of sowing, uniformity in spacing and above all rapid coverage that it ensures. At the same time it makes post-sowing operations such as weeding, interculture, etc., more convenient and effective. Use of seed-cum fertilizer drill helps to attend two operations simultaneously, ensuring optimum placement of seed and fertilizer in moist zone and enabling higher fertilizer use efficiency.

(ii) **Dry Seeding:** Sowing of crops immediately after the receipt of soaking rains is the traditional practice of rainfed farmers which may be called monsoon or wet sowing. But in vertisols where dependable rains followed by reliable monsoon rains exist, sowing of bold seedbed crops before the onset of monsoon rains in dry soil is advantageous and confirms multiple benefits. Since here even the first drop of monsoon rain itself is used for crop germination, growing period is lengthened as against taking up wet sowing which can be done only with considerable loss of moisture in the event of first rains happen to be heavy and incessant making the soil too sticky thus postponing the seeding unduly. The sowing period is, therefore, shortened. Dry seeding in anticipation of rains helps to establish a crop at an early opportunity and gives an advantage of better utilization of rainfall during crop growth. Dry seeding is found beneficial in sorghum, castor, cotton, pigeonpea, etc.

Soil necessarily should be dry for dry sowing otherwise seeds get damaged under semi wet conditions leading to poor stand establishment by insects and pathogens. To avoid damages to seed, seed pelleting with protective chemicals could be followed. After uniform coating, dry the seeds in shade, open or sun drying results in cracking and removal of coating. Treated seeds remain in soil unaffected for 20–25 d or more. Sowing is to be done based on rainfall probability. Alternatively, double seed rate can also be followed to make up any seed loss where seed cost is quite low.

(iii) **Deep Sowing:** Depth of seeding has been observed to be critical for the establishment of satisfactory plant stand. During rainy season a sowing depth of 2–3 cm is optimum in case of pearlmillet, fingermillet, other millets, and sesamum. For maize, sorghum, wheat, barley, pulse crops and most of the oilseeds, 5–6 cm seeding depth has been found optimum. Deep sowing particularly in post-rainy season cropping or stored soil moisture is necessary for crops having long coleoptiles lengths like durum wheat, sorghum, chickpea, etc. This practice helps to place seed in moist zone for successful establishment and which in later stages of growth can withstand water stress to a greater degree due to well developed roots into the soil.

(iv) **Seed Hardening:** Pre-sowing hardening of seeds is one of the methods of modifying the physiological and biochemical nature of seeds so as to get the characters that are favourable for drought tolerance. Drought tolerance of many plants can be increased by subjecting them for several days to moderate dehydration, low temperatures or treatment with growth retarding substances. Seed hardening is, therefore, done to overcome the adversity of moisture stress that is common in dry farming areas. When dry seeds are soaked in water/chemical solution, the quiescent cells get hydrated and germination initiated. It also results in enhanced mitochondrial activity leading to the formation of high energy compounds and vital bio-molecules. The latent embryo gets enlarged. When the imbibed seeds are dried again, the triggered germination is halted. When such seeds are sown, reimbibition begins and the germination event continues from where it stopped previously. As a consequence there will be multitudes of benefits, which make the plants prepared to resist the adverse weather if any. Beneficial effects of seed hardening includes, accelerated rapid germination and growth rate of seedings. Plants recover much more quickly from wilting, resistance to salinity as well as to drought condition, flowering is slightly accelerated, early emergence, enable to compete more effectively with weeds and ultimately results in more yield.

Seed hardening involves seed soaking in chemical solution/water for a specified period (6–16 h) at room temperature followed by shade drying for 24 h. Some of the chemicals recommended for different crops are given as under in table - 6.4.

Table.6.4: Chemicals Recommended for Seed Hardening in Different Crops.

Crop	Chemical
Maize	Cycocel (250 ppm)
Sorghum	KH_2PO_4 (2%) or $CaCl_2$ (2%)
Pearlmillet	NaCl (0.5%)
Paddy	KCI (1%)
Millets & cotton	KCI (2%)
Groundnut	$CaCl_2$ (0.5%)
Pigeonpea	$ZnSO_4$ (100 ppm)
Sunflower	GA_3 (500 ppm) or $MgSO_4$ (0.2 – 0.5%)
Greengram	$MgSO_4$ (100 ppm)
Chickpea	$KH_2 PO_4$ (1%)

6.5 Optimum Plant Densities and Crop Geometry

Optimum plant population is a pre-requisite for higher yields. Higher plant population than the recommended creates more interspecifc competition resulting in reduced yield. In sorghum, maize and other crop plants even remain barren when density is more than optimum. Number of plants per unit area largely depends on available soil moisture. In branching crop like safflower, plants show considerable adjustments to plant density and spacing by regulating branching. At low density plants produce higher number of secondary and tertiary branches compensating the low density. On the contrary high plant density produces less secondary branches. Thus over a range of population yield remain at the same level. The traditional practice of planting with maximum population pressure in drylands does not help. In fact it may work negatively by producing barren plants. The recommendation on plant densities in dryland conditions should be therefore, based on the study of the local situations. Recommended crop geometry for different crops is given in table 6.5 as under.

Table 6.5: Crop Geometry for Dryland Crops.

Crop	**Crop geometry (R × R & P × P spacing in cm)**
Maize	60 × 25 (long duration); 45 × 30 (short duration)
Sorghum	45 × 15
Pearlmillet	45 × 10-12
Small millets	30 × 10
Blackgram, greengram, horsegram	30 × 10
Pigeonpea	60 × 30
Cowpea	30 × 15
Soybean	30 – 45 × 10–15
Groundnut	30 – 45 × 10–15
Sesame	30 – 35 × 10–15
Sunflower	30 × 15
Castor	50 – 60× 45
Cotton (American)	60 × 45
Cotton (Desi)	60 × 30
Wheat	20 – 23 (R × R)
Barely	25 (R × R)
Chickpea	30 – 45 (R × R)
Mustard	30 – 45 × 10
Taramira	30 × 10
R = Row and P = Plant	

6.6 Intergrated Nutrient Management

Application of plant nutrients in dry land crops is mainly governed by soil type, quantity of rainfall and its distribution and crops grown. Applied level of fertilizer to dryland crops is almost half of the recommended fertilizer levels in irrigated agriculture. Recommended dose of nutrients to different dryland crops is given as under in table 6.6.

Table 6.6: Fertilizer Recommendations for Dryland Crops.

Crops	Recommended major nutrients (kg ha^{-1})		
	Nitrogen (N)	Phosphorus (P_2O_5)	Potash (K_2O)
Maize	40 – 60	30–40	20
Sorghum	50 –60	20 – 30	20
Pearlmillet	50 – 60	20–30	20
Up land rice	40 – 60	20 – 30	20 – 25
Minor millets	20 – 40	20	—
Kh. Pulses	10 – 15	20 – 25	—
Pigeonpea	10 – 15	20 – 30	—
Soybean	15 – 20	20 – 30	20 – 30
Groundnut	10 – 15	25 – 30	15–20
Sesame	15 – 20	10 – 15	—
Castor	30 – 40	20 – 25	20
Sunflower	30 – 40	30	20
Cotton	40 – 50	20 – 25	—
Wheat	50 – 60	20–25	20
Barely	25 – 30	15 – 20	—
Chickpea	10 – 15	20 – 30	—
Rapeseed & mustard	30 – 45	30	20
Taramira	20 – 25	15 – 20	—
Safflower	15 – 20	15	10

* The dose may vary for different crops in different regions of the country.
* If soils are deficient in zinc or iron it may be applied.

Full dose of phosphorus and potash and half dose of nitrogen are applied as basal in general except pulses where full dose of fertilizers is applied at sowing only. The remaining half quantity of nitrogenous fertilizer may be applied as top dressing in one or two times after 30-35 days of sowing depending upon the type and stage of the crop with availability of rainfall or moisture in the soil. The possibility of foliar nutrition of nitrogenous fertilizer can be looked into if soil moisture is not adequate for field application of fertilizer.

Use of organic manures like FYM, compost, vermicompost, bio-gas slurry, etc. are important for dryland conditions. These bulky organic manures release the essential nutrients slowly which favours the growth of plants in addition to improvement in physical properties especially organic matter content, water holding capacity, soil air, soil temperature, etc. They also act as an important source of food for microbes in soil, which improve the availability of nutrients. In general 10-12 tonnes of FYM / compost or 5-6 tonnes of vermicompost or biogas slurry is recommended per hectare in dryland areas atleast once in 2-3 yr.

Bio-fertilizers are eco-friendly, environmentally safe and cost effective source of nutrients that are best suited under dryland conditions. *Rhizobium*, phosphate solubilizing bacteria, *Azospirillum, Azotobacter* and *Vasicular arbuscular mycorrhizea* (VAM) are important bio-fertilizers, which can be effectively used under dryland conditions. Bio-fertilizers recommendations for different crops are given as under in table 6.7.

Table 6.7: Bio-fertilizer Recommendations for Different Crops.

Crop	Bio-fertilizer	Quantity (kg ha^{-1})	Method of application
Upland rice	*Azospirillum*	2.0	To be mixed with 25 kg each of FYM and sand and broadcast before sowing.
	Azospirillum	0.6	Seed treatment
Sorghum	*Azospirillum / Azotobacter*	0.6	Seed treatment
	Phosphobacterin	0.6	Seed treatment
	Azospirillum	2.0	Soil application with FYM & sand as above
	Phosphobacterin	2.0	- do -
Maize	*Azospirillum / Azotobacter*	2.0	Soil application with FYM & sand as above
	Phosphobacterin	2.0	- do -
Pulses	*Rhizobium* multi strain	0.6	Seed treatment
	Phosphobacterin	0.6	Seed treatment
Soybean	Brady *Rhizobium japonicum*	0.6	Seed treatment
	Phosphobacterin	0.6	Seed treatment
Groundnut	*Rhizobium &* Phosphobacterin	0.6	Seed treatment
	Phosphobacterin	2.0	Soil application with FYM & sand as above
Cotton	*Azospirillum*	0.6	Seed treatment
		2.0	Soil application with FYM & sand as above

In addition to above there are other ways to supply nutrients in dryland crops are inclusion of legumes in cropping sequence, intercropping, green manuring, etc.

6.7 Intergrated Weed Management

Weeds in general reduce crop yields by about 40 per cent in rainy season. Since farmers adopt some kind of weed management on their fields, conservative estimate of about 15 per cent yield loss in dryland crop may be taken as more realistic. Competition for moisture becomes critical with increasing soil moisture stress as found in arid and semi-arid areas. In general the initial 4 to 6 wk period is more critical from the weed crop competition point of view in almost all the crops. The semi-arid and arid environments being ecologically fragile enclosed with poor resource base of farmers, weed management option should be effective and economically viable to minimize drudgery and to be environmentally neutral or favourable.

Off season tillage helps to minimize weed problem in *kharif* crops on light soils. During cropping weeding time is important. First weeding has to be necessary, to be completed within 20–25 d after sowing. Manual method of weed control is the oldest and simplest method of removing weeds within rows and hills of crops where weeds cannot be controlled by mechanical means. The efficiency of manual weeding could be increased using various hand tools such as hoe, *khurpi,* spade and double blade knives depending upon the crop, spacing and density of weeds. Use of power weeders is a better option as it saves time and cost as well. For all mechanical weed control methods line sowing is pre requisite. Similarly stale seed bed technique involves primary cultivation followed by a period in which seeds are induced to germinate and they are destroyed by further cultivation or herbicide use before sowing of the crops. This technique is of great importance in rainfed farming where moisture regime dictates weeding during critical cropping period. Crop rotation also plays a significant role in the management of certain host specific parasitic weeds for example cotton, sunflower; cowpea, groundnut and castor are important trap crops of *Striga* spp, a parasitic weed in sorghum and pearlmillet. Ecological suppression of weeds is also possible by intercropping systems. This phenomenon is particularly important for smother crops or living mulch system in which low growing weed suppressive species are grown between tall growing crops.

Use of herbicides to control weeds in dryland agriculture is the best option to check the early weed crop competition as due to continuous rains or scarce availability of labours timely weed control becomes difficult. The selective herbicides for different crops are given in table-6.8.

Table 6.8: Selective Herbicides for Weed Control in Major Dryland Crops

Crop	Herbicides	Dose	Time of spray
Cereals			
Rice	Anilophos	0.4–0.5 kg ha^{-1}	PE at 6–7 DAS
	Oxyfluorfen	0.15–0.25 kg ha^{-1}	PE at 0–6 DAS
	Bensulfuron	30–50 g ha^{-1}	PE at 3–5 DAS
	Chlorimuron + mestsulfuron methyl	4 g ha^{-1}	POE at 15–20 DAS
Maize	Alachlor	2–2.25 kg ha^{-1}	PE at 0–3 DAS
	Tembotrine	100 g ha^{-1}	POE at 15–20 DAS
Sorghum	Alachlor	1.0-1.5 kg ha^{-1}	PE at 0-3 DAS
Oilseeds			
Groundnut	Alachlor	1.5 kg ha^{-1}	PE at 0–3 DAS
	Nitrofen	1.5 kg ha^{-1}	PE at 0–3 DAS
	Oxyfluorfen	0.25 kg ha^{-1}	PE at 0–3 DAS
	Quizalofop ethyl	40–50 g ha^{-1}	POE at 15–20 DAS (only for grassy weeds)
	Imazethapyr	100 g ha^{-1}	POE at 10–15 DAS
Rapeseed & mustard	Oxadaizon	0.5–0.75 kg ha^{-1}	PE or POE
	Quizalofop ethyl	40–50 g ha^{-1}	POE at 15–20 DAS (only for grassy weeds)
Soybean	Alachlor	1.5–2.0 kg ha^{-1}	PE at 0–3 DAS
	Imazethapyr	100 g ha^{-1}	POE 10–15 DAS
	Oxyfluorfen	0.1–0.25 kg ha^{-1}	PE at 0–3 DAS
	Quizalofop ethyl + chlorimuron	30–40 g ha^{-1}	POE at 20–25 DAS
Sesame	Oxadizon	0.5 kg ha^{-1}	PE or early POE
Pulses			
Greengram & blackgram	Alachlor	2.0–2.5 kg ha^{-1}	PE at 0–3 DAS
	Oxyfluorfen	100–125 g ha^{-1}	PE 0–3 DAS
	Quizalofop ethyl	40– 50 g ha^{-1}	POE 15–20 DAS (only for grassy weedy)
Pigeonpea	Alachlor	2.0–2.5 kg ha^{-1}	PE at 0–3 DAS
	Oxyfluorfen	100–125 g ha^{-1}	PE at 0–3 DAS
	Pendimethalin	0.75-1.0 kgha^{-1}	PE at 0-3 DAS
Chickpea	Metolachlor	1.0–1.5 kg ha^{-1}	PE at 0–3 DAS
	Quizalofop ethyl	40–50 g ha^{-1}	POE at 15–20 DAS (only for grassy weeds)

PE– Preemergence, **POE**– Post- emergence, **DAS**– Days After Sowing

◆

7

CONTINGENT CROP PLANNING FOR ABERRANT WEATHER CONDITIONS

Unfavourable deviated weather from the normally expected one is called **aberrant weather.** Aberrant weather is a common feature of rainfed agriculture. Crop production in dryland areas is mainly dependent upon rainfall, which is received during the south west monsoon. The rainfall is seasonal, erratic and highly variable with space and time. Vagaries of monsoon rains and breaks of varying duration are most common in India. Aberrant weather conditions appears in the form of delayed onset of monsoon, dry spell immediately after sowing, short or long dry spells during growing season *i.e.* in mid season or late season (break in monsoon-mid or late season drought) and early withdrawal of monsoon.

Contingent Crop Planning

Contingent crop planning means to mitigate any unexpected, unusual, unfavourable weather situations occurring at any time without prior knowledge at any time before the crops are sown or even after crops are sown. The contingent crop planning therefore is proposed to mitigate such situation through the choice of appropriate crop and varieties, cropping systems and other necessary relevant farm practices. Contingent plans have been developed by Central Research Institute for Dryland Agriculture (CRIDA), Hyderabad for different districts of the country based on the research available in different agro-climatic zones. Adoption of these plans may be helpful to stabilize production in these areas. Contingent crop plans for different aberrant weather situations is given as under:

7.1. Contingent Crop Plan for Different Aberrant Weather Conditions

7.1.1 Delayed Onset of Monsoon

During some of the years the onset of south west monsoon gets delayed by 20–40 d. Such a late onset shortens the length of growing season, restricts the choice of crops and varieties, limits use of inputs, results in low yield of *kharif* crops due to delayed sowing and leaves less moisture for post rainy season crops grown in succession under these conditions. Management options available in this condition are as follows:

(i) **Dry Seeding:** In delayed monsoon conditions dry seeding is beneficial as every drop of rainfall can be used for crop germination, growing period is lengthened as against taking up wet sowing which can be done only with considerable loss of moisture in the event of brief rains happen to be heavy and incessant making the soil too sticky thus postponing the seeding unduly. The sowing period is, therefore, shortened. Dry seeding in anticipation of rains helps to establish a crop at an early opportunity and gives an advantage of better utilization of rainfall during crop growth. Dry seeding is found beneficial in bold seeded crops like maize, soybean, castor, cotton, pigeonpea, etc. For dry seeding soil should be necessarily dry, otherwise seeds get damaged under semi wet conditions. There may also be the problem of insects particularly of ants and other pathogens. To avoid damages to seeds, seed pelleting with appropriate chemicals could be followed. Avoid to use hybrid/costly seeds.

(ii) **Raising Nursery and Translating:** In crops like upland rice, fingermillet, pearlmillet, etc, wherever transplanting is possible nursery may be raised for sowing in delayed onset conditions. This is a surer way of compensating the delay of monsoon, though transplanting is a labour intensive operation.

(iii) **Alternate Crops and Varieties:** Crops /varieties differ in their capacity to produce better in moisture deficit conditions and even if sown late due to their shorter duration or deep root system, etc. depending on the receipts of rains, such crops and varieties can be selected depending on date of occurrence of rains. Planting of maize for grain after 15th of July is not advisable. Planting of pearlmillet is not recommended beyond July even short duration hybrids. In soybean if monsoon is delayed by 2–3 wk, short duration cultivars are preferred. In upland rice, if monsoon is delayed, short duration upland rice varieties suitable for direct seeding or transplanting with older nurseries, if varieties are suitable is suggested. Beyond July it is preferred to do sowing of short duration pulses like blackgram, greengram or oilseeds like castor, sesame or niger or maize /sorghum/ pearlmillet for fodder. In delayed onset of monsoon it is always preferred to go for short duration early maturing varieties of crops recommended for different agro-climatic zones of the country.

7.1.2 Dry Spell Immediately After Sowing

During some of the years dry spell may occur immediately after sowing of the crop. This may result in poor germination due to crust formation and poor establishment of crop stand due to inadequacy of soil moisture at sowing. The suggested measures for this weather condition are as under:

(i) **Resowing:** When there is a severe dry spell and the population is too poor to expect any reasonable yield, resowing of the entire field using same or new crop using subsequent rains is necessary.

(ii) **Gap filling:** If the dry spell is soon after sowing and there is sub-optimum plant stand with the subsequent showers, gap filling may be considered advantageous. If the sown crop is suitable for transplanting and the stage of the seedlings is appropriate, transplanting in the gaps using excess seedlings growing at some places in the crop field with subsequent rains. If the break in rains is rather long and do not permit transplanting of already sown crop with existing seedlings due to age, other crops grown in the region can also be used for transplanting e.g. pearlmillet, fingermillet, vegetables, etc. Similarly re-seeding can also be done in the gaps using other crops suited for late planting.

7.1.3 Break in Monsoon –Mid or Late Season Drought

In dry lands moisture is insufficient for crop production quite often. Farmers of dryland areas take up sowing when the soil profile is wet up to 30–40 cm depth because it is sufficient to meet initial crop requirement of the moisture. Plants tolerate stress up to one month and later on start wilting. During this stage there is need to introduce or super impose mid season correction to save standing crop otherwise whole crop may be lost. Several practices have been developed and suggested to mitigate the damaging effects of dry spells as follows:

(i) **Thinning of the Stands:** Thinning of the plant stand is recommended to minimize the severity of moisture stress of about 30–50 d. Thinning of alternate plants in the rows in the event of mild stress and every third row under higher moisture stress condition is recommended. Thinning is resorted particularly to the most sensitive component for soil moisture such as sorghum in sorghum + pigeonpea intercropping system.

(ii) **Removal of Weak Seedlings:** In most of the dryland crops there may not be uniform emergence because of soil variation, differential depth of sowing or inadequacy of soil moisture at sowing. In depressions there may be quick germination and on ridges there may be slow germination. In most of the cereals, the late germinated seedlings will not be vigorously growing and may not be competitive to give any yield. At severe drought conditions these weak seedlings may be removed.

(iii) **Repeated and Deep Intercultivation:** Early breaks in rainfall bring about soil cracking early in the season which enhances evaporation loss of moisture through the cavities. To prevent soil cracking and to break the capillary loss, early and frequent inter-cultivations are needed. Generally blade hoes are employed for the purpose which penetrate hardly to a depth of 5 to 7.5 cm. Shovel type of interculturing hoes with iron coulters should be used which due to sharp cutting edge can tear off the soil and create a dust mulch to a depth of 10–12 cm.

(iv) **Mulching:** Mulches have a definite controlling effect on the evaporation of soil moisture and hence play an important role in dryland agriculture. Under field conditions, surface soil is just stirred and the soil becomes air–dry. By adopting this practice, some amount of moisture is lost from the upper layer. After the preliminary loss of moisture, the soil mulch effectively controls further evaporation from deeper layers by breaking the capillary continuity. This system is highly suitable for medium to heavy textured soils.

(v) **Stripping of Older Leaves:** Older leaves are shaded, they transpire more than helping in photosynthesis. So to avoid moisture loss these older leaves may be stripped, which helps in active growth of new parts. For example in sorghum 12–15 leaves are produced, when the plants are at flag leaf stage strip off the bottom 3–4 leaves and use them for mulching or for feeding to the livestock . Similarly in maize growing areas it is a practice to cut the top portion above the cob after milky stage.

(vi) **Returning Ratooning:** In case of pearlmillet and sorghum when crop is exposed to dry spells of 30–35 d during peak growth (blooming stage), yields are drastically reduced. Under such conditions cut / harvest the crop for fodder purposes and allow the ratoon with the onset of rains (after experiencing drought). For quick regrowth supplemental addition of fertilizer is suggested.

(vii) **Life Saving Irrigation:** If there are farm ponds/seasonal streams, life saving irrigation should be given to save the crop. Runoff is inevitable in the tropical and sub-tropical climates of the country.

(viii) **Use of Anti–transpirants:** The possibility of reducing transpiration is of greatest importance in dry farming. Many herbicides, fungicides and growth hormones posses anti–transpirant properties may be of stomatal closing type or film forming type or reflectant type or growth retardant. Some of the fungicides like phenyl mercuric acetate (PMA) and herbicides like atrazine in low concentrations serve as anti-transpirant by inducing stomatal closing. The plastic and waxy materials which form a thin film on the leaf surface retard the escape of water due to formation of a physical barrier e.g. mobileaf, hexadecanol, silicon, etc. The success of these chemicals is limited since they also reduce photosynthesis. Some of the white materials form a coating on the leaves and increase the leaf reflectance (albedo). Transpiration loss is reduced by reflecting the radiation reducing leaf temperature and vapour pressure gradient. Spraying of kaolin (5%), a reflectant type of antitranspirant, forms thin layer on the leaf growth checks transpiration losses. It is low cost anti–transpirant and easy to use by the farmers. Growth retardants such as cycocel induce stomatal closure and thus reduce transpiration losses from leaf surface.

(ix) **Harvest the Crop for Fodder:** If there is complete failure of rains and crop is not in position to produce seed yield especially the *kharif* crops, in a double cropped area, such crop may be harvested for fodder e.g. sorghum, maize, pearlmillet and the land could be kept ready to make use of shower for post rainy season crops like chickpea, mustard, toria, taramira, linseed, lentil, etc.

7.1.4 Early Withdrawal of Monsoon

If monsoon terminates earlier, then it cuts the length of growing season. Normally crops in their maturity stage and those growing on shallow soils suffer heavily. In this condition sowing of *rabi* crops may also get affected. It is therefore, suggested to use short duration and early maturing varieties to face the situation in well identified dryland areas.

7.2 Specific Contingent Practices for Different Crops/ Crop Groups

7.2.1 Pearlmillet/ Small Millets

- Planting of pearlmillet hybrids is not advisable. Even early maturing hybrids like HHB 67 should not be planted beyond July.
- Wherever crop has been planted and is suffering for moisture stress, the plant population may be reduced and shallow interculture (dust mulching) may be practiced.
- The pearlmillet may be grown mixed with pulses up to first week of August as fodder crop, which may give atleast fodder to some extent.
- In years of extreme drought and erratic monsoon conditions small millets like foxtail millet for Rajasthan, Kodo, proso and little millets for Gujarat and little millet (Kutki) and Kodo millets for M.P. could be ideal crops for contingency planning for producing quickly fodder and grain.

7.2.2 Sorghum

- In case there is delay in monsoon by 2–3 wk short duration varieties/ hybrids may be preferred.
- In staggered/delayed planting increase seed rate to an extent of 1.5 times and apply recommended insecticides in the seed rows before sowing to safeguard the crop against the anticipated shoot–fly attack.

7.2.3 Upland Rice

- In traditionally rice growing rainfed areas where rains are likely to come late and where a normal transplanted rice crop is ruled out, short duration upland rice varieties or those rice varieties that are suitable for direct seeding either in dry or wet conditions and subsequent flooding are recommended.
- In certain areas, delayed transplanting can be done with older nurseries if these varieties are suitable.

7.2.4 Maize

- Planting of maize for grain after 15th July is not advisable.
- Intercropping of blackgram or pigeonpea with maize is better under dryland conditions.
- After 15th July plant maize only for fodder purpose.

7.2.5 Pulses

- In the event of late monsoon arrival the pulse crops which can be planted preferably up to 15th August or till 1st week of September are: blackgram, greengram, pigeonpea and kidneybean.
- Only specific varieties of pigeonpea may be grown up to first week of September which are recommended for pre-*rabi* planting. These varieties are Bahar, Pusa-9 and *rabi* arhar 20(5) in the states of Bihar, West Bengal, Orissa, Andhra Pradesh, Maharashtra and Gujarat. Planting of early pigeonpea is not possible in states like Punjab, Haryana and western U.P. where early maturing varieties of pigeonpea normally grown in first week of July.
- Crops like kidneybean (mothbean) can be grown for grain as well as fodder till the second week of August in states of Rajasthan and Gujarat.

7.2.6 Oilseeds

- In case of oilseed crops, sesame can be planted up to third week of August whereas niger can be sown up to September in some parts of the country. Since the per hectare requirement of seed in case of sesame is very low, all out efforts must be made to cover more area under sesame to utilize the land in the event of failure of other *kharif* crops like maize, pearlmillet and other minor millets.
- Similarly, castor can be grown successfully up to first week of September in areas, which are suitable for castor cultivation.
- The other two important oilseed crops like soybean and groundnut are likely to suffer heavily in case rains are delayed beyond July end. However, groundnut can be planted up to September in a few southern states. Therefore, efforts should be made to cover more area under groundnut in the states of Andhra Pradesh, Tamil Nadu and Karnataka.
- In delayed rainfall conditions, there seems good possibility for toria in the states of Punjab, Haryana, Madhya Pradesh, Uttar Pradesh , Bihar and West Bengal. All recommended varieties of toria can be promoted as a *rabi* crop.

7.2.7 Cotton

- Use of micro-irrigation system such as drip and sprinkler may be advocated for giving life saving irrigations, if harvested water is available.

- Development of ridge and furrow across the slope for effective conservation of soil moisture as well as rainwater.
- Use of organic mulches such as subabul loppings, straw, etc. to conserve the soil moisture.
- Repeated intercultural operation to keep the field weed-free.

7.3 Contingent Plan Relevant to all Crops and Cropping Systems

7.3.1 Special Package of Practices

- Making fields free of weeds for full utilization of water and nutrients by the crops.
- Reduction of plant density in case of mid season drought. The crops should be suitably thinned out or ratooned in case of sorghum. In case drought occurs at very early stage, it is always better to resow with subsequent rain rather than allowing sub-optimal poor plant stand to persist.
- For drought conditions anticipating prolonged dry spell/breaks, the practice of intercropping helps in risk distribution. This can be achieved by including a companion crop like blackgram, greengram, cowpea with the main crops like maize, sorghum and pearlmillet.
- The recommended dose of nitrogen application should be reduced by 40–50 per cent under unirrgated conditions and should always be applied at sowing/planting. In standing crop top dressing of N can be done in cereals and millets as per need of the crop and soil moisture conditions.
- Full recommended dose of phosphorus and potash should be placed as basal dose.
- Resorting to maximum possible use of organic bulky manures.
- Recommended plant protection measures to be adopted for *kharif* millets and pulses.
- Seeds of sorghum, pigeonpea, greengram, cowpea, groundnut, sunflower and castor must be treated with proper fungicides as per recommendations.

7.3.2 Soil and Water Management Practices

- Seed beds to be kept ready so as to facilitate sowing immediately with the onset of rains. Field should be properly levelled for uniform water distribution within the sub-plot.
- Broad bed and furrow, ridge and furrow, compartmental bunding and contour trench land configuration may be adopted in shallow alfisols as moisture conservation practices.

- Furrow sowing of *kharif* crops at closure plant to plant spacing with wider inter row spacing.
- Frequent inter culture to facilitate effect of loose soil as dust mulch. Wherever economically viable, mulching should be practiced in between crop rows using locally available mulching material.
- Wherever possible run-off may be harvested to make provision for protective irrigation at later stage/crop ripening.
- Major emphasis should be on *in-situ* rainwater conservation, harvesting excess run-off for re-use and ground water recharge.

◆

8

WATERSHED MANAGEMENT

Land, water and vegetation are the most important natural resources for survival of the man and livestock. To obtain the optimum production of bio-mass these natural resources have to be managed efficiently. As these resources are inter dependent, there is a need to have a unit for most effective and useful management of these resources. In this context, watershed is an important unit for the management of the natural resources. The word watershed has different meanings. In British English, it means a ridgeline or a line which shows slopes in two different directions on its either side. A ridgeline is also a line connecting the points of highest elevation in a terrain. Therefore, ridgeline is also known as "watershed" line, or a surface water divide. In American English, the word watershed is used as a synonym for "catchment" or "basin" wherein rainwater or storm water gets collected from an area, enclosed by a ridgeline. This water eventually flows through the various drainage channels which merge with one another to form one or rarely more than one outfall(s) of a stream. Thus watershed is defined as an area enclosed within a watershed line. In this course, the word basin or a small catchment representing a hydrological unit, which drains all its rain water into a stream. Watershed areas of the country are suffering from various problems like degradation of land, improper land use, excessive cropping, shifting cultivation, over grazing, human and animal pressure, etc. Therefore, to optimize the integrated use of land, water and vegetation in an area to alleviate drought, moderate floods, prevent soil erosion, improved water availability and increase food, fodder, fuel, fibre on sustained basis in dryland areas watershed management programme is adopted.

8.1 Definition of Watershed

Various definitions have been proposed over the recent past for the term **"watershed".** Since large watersheds consist of many smaller watersheds, it is necessary to define the watershed in terms of a point. This point is usually referred to as the watershed **"outlet".** With respect to the outlet, the watershed consists of all land area that sheds water to the outlet during a rainstorm.

A watershed is a topographically delineated area that is drained by a stream system, *i.e.* the total land area that is drained to some point on a stream or river. A watershed is a hydrological unit that has been described and used as a physical biological unit and also, on many occasions, as a socio-economic political unit for planning and management of natural resources.

A watershed, also known as drainage basin or catchment area is an area of land where all water drains to a control point like, river or stream. Another way watershed can be defined as a hydrological unit where from surplus runoff drains through a common outlet. The watershed boundary is called **drainage divide.** Rains received on opposite sides of drainage divide do not contribute runoff and reach to the adjoining areas.

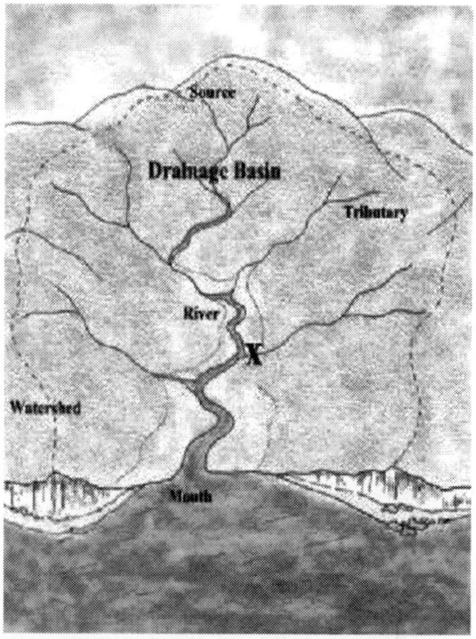

Fig. 8.1: Sketch of Watershed.

Watershed is also defined as any surface area from which runoff resulting from rainfall is collected and drained through a common point. It is synonymous with a drainage basin or catchment area. Watershed may be only a few hectares as in small ponds or hundreds of square kilometers as in rivers. A basic sketch of watershed is given in fig. 8.1.

8.2 Watershed Management

Watershed management is a term used to describe the process of implanting land use practices and water management practices to protect and improve the quality of water and other natural resources within the watershed by managing the use of those land and water resources in a comprehensive manner. Watershed management is an adaptive comprehensive integrated multi resource management planning process that seeks to balance healthy, ecological, economic and cultural/social conditions within a watershed. Water management serves to integrate planning for land and water, it takes into account both ground and surface water flow, recognizing and planning for the interaction of water, plants, animals and human, land use found within the physical boundaries of a watershed.

Watershed management implies the rational utilization of land and water resources for optimum production with minimum hazards to the environment. To achieve this adoption of soil and water conservation practices are essential in the watershed. The aims of these practices are proper land use, protecting land against all forms of deterioration, building and maintaining soil fertility, conserving water for farm use, proper management of local water for drainage, flood protection, sediment reduction and increasing productivity from all land use.

It is important that watershed management must consider the social, economic and institutional factors operating within and outside the watershed area. All watersheds contain various types of natural resources. The key is to use these resources efficiently as far as possible with minimum watershed degradation. Since watershed management involves decision making for efficient utilization of natural resources, a multidisciplinary approach is essential. In general, watershed based planning of resource management has generated wide appreciation, particularly for assured dividends. Recently, concept of maintaining an ecological balance embedded in the watershed programme has also started getting major attention in different sections of the society.

8.3 Concept of Watershed Management

Thus, watershed management is an approach of area planning of natural resources, specially land, water and plants to serve the socio-economic needs of the community concerned. It has to be sustainable on ecosystems principles when the treated watershed would permit maximum possible stability through the processes of production, consumption and regeneration. It has to be sustainable in terms of meeting substance and trade off needs of the household or livelihood of its members over the years to come. For land development purposes, a convenient size of watershed is 500–1000 ha, but these limits not to be taken rigidly.

As the entire process of agricultural development depends on the status of water resources, the watershed with the distinct hydrological boundary is considered ideal for taking development programme. In brief, planning and design of soil and water conservation structures such as bunds, waterways, overflow hydraulic structure, water harvesting structures, etc. are carried out considering the expected rate and amount of runoff water. In this context, the watershed concept is of practical significance in estimating the rate of runoff and flood volumes to be handled by these structures. Thus, it is essential to have various developmental programmes on watershed basis as conservation measures or hardware treatment. And to be on safer side, the entire development needs to be taken up on topographic considerations from the ridgeline to the outlet point. The people and animals are also part of the watershed community. All depend on the watershed and they in turn influence what happens there whether it is good or bad. What happens in a small watershed also affects the larger watersheds.

Development is not confined to agricultural lands alone but covers the entire watershed area, starting from the highest point or ridgeline to the outlet of the nallah or natural stream. In fact, the sequence of land development is always more efficient if it starts at the ridgeline. This involves treatment of barren hill slopes, marginal lands, privately owned lands, nallahs and river courses in that order.

Further concept of watershed can also be defined as the application of business method and technical principles to the manipulation and control of watershed resources to achieve a desired set of objectives such as maximum supply of usable water, minimization of erosion and siltation problems and reduction of flood and drought occurrence.

8.4 Watershed Approach

A watershed is a geo-hydrological unit, which drains into a common point. The watershed approach is a project based ridge to valley approach for in-situ soil and water conservation, forestation, etc. Unit of development will be a watershed area of about 500 ha each in watershed development projects. However, the actual area of a project may vary keeping in view the geographical location, the size of village, etc. The thematic maps generated from satellite data for different themes such as land use/land cover, hydro-geomorphology, soils, etc. may be used for selection of a watershed area. The project will primarily aim at treatment of non-forest wastelands and identified drought prone and desert areas. However, if any watershed area consists of some forestlands, it should also be treated simultaneously under the project.

The **size** of a **watershed** can vary depending on the point of discharge *chosen* to define it. If the chosen point was the mouth of a major river that drains to the ocean, the **watershed** would likely be very large. However, if you chose a point up river at the confluence of two streams, the **watershed** would be much smaller based on the geographical area. Watersheds could be classified into different categories as given in table 8.1.

Table 8.1: Classification of Watershed on the Basis of Area

Type	Area (ha)
(i) Macro- watershed	>50,000
(ii) Sub-watershed	10,000-50,000
(iii) Milli-watershed	1,000-10,000
(iv) Micro-watershed	100-1,000
(v) Mini-watershed	1-100

8.5 Principles of Watershed Management

The main principles of watershed management are as follows:

- Utilizing the land according to its capability class.
- Maintaining adequate vegetative cover on the soil for controlling soil erosion.
- Conserving maximum possible rainwater at the place where it falls; on arable land, by contour farming, etc.
- Draining out excess water with a safe velocity to avoid soil erosion, and storing it in ponds for future use.
- Preventing erosion in gullies and increasing groundwater recharge by putting nallah bunds and gully plugs at suitable intervals.

- Ensuring sustainability of the ecosystem benefiting the man-animal-plant-land-water complex in the watershed.
- Maximizing productivity of all the existing resources.

8.6 Objectives of Watershed Management

The main objective of watershed management is the proper use of all the available resources of a watershed for optimizing productivity with minimum hazards to natural resources. The other objectives of watershed management are–

- Recognition of watershed as a unit for development and efficient use of land according to their land capabilities for production without any environmental degradation;
- Conservation, development and sustainable management of natural resources including their use;
- Conserve as much rain water as possible in the place where it falls, of course without disturbing the natural hydrological system, increasing the groundwater level and maintaining it for sustainable use;
- Prevent soil erosion and reduce sediment production by means of suitable soil and water conservation measures;
- Restoration of ecological balance in the degraded and fragile rainfed ecosystem by greening these areas through appropriate mix of vegetation, viz, trees, shrubs and grasses.
- Enhancement of agricultural productivity and production in sustainable manner.
- Reduction in regional disparity between irrigated and rainfed areas.
- Improve infrastructural facilities in the watershed.
- Provide training and build capacity of the primary stakeholders through participatory approaches so as to evolve a demand driven action plan for sustainable watershed development.
- Creation of sustained employment opportunities for the rural community including the landless, and
- Evolve monitoring and evaluation methods to make the programme transparent, flexible and cost effective. Community based organizations and the people shall be partner in this endeavour.

8.7 Benefits of Watershed Management

- Sound watershed management means controlling floods and reducing erosion as well as sediment accumulation.
- Maximizing productivity per unit area, per unit time and per unit of water.
- Increasing cropping intensity.

- Proper utilization of marginal or wasteland through alternate land use systems.
- Ensuring ecological balance.
- Maximizing the combined income from the inter related crop, livestock, tree, labour complex over the years.
- Stabilizing income even under unfavourable weather conditions and cutting down risks during aberrant weather conditions.

8.8 Components of Watershed Management Programme

The main components of watershed programme are; soil and water conservation, water harvesting, crop management and alternate land use system as discussed below:

8.8.1 Soil and Water Conservation & Water Harvesting

Soil and water conservation measures coupled with water harvesting help to improve the moisture availability in the soil profile and surface water availability for supplemental life saving and off season irrigation. Based on the barriers and their cost, the conservation measures in arable lands can be divided in three categories namely hardware, medium software and software treatments. The details of these treatments are given as under –

8.8.1.1 Soil and Water Conservation in Arable Lands

(a) Hardware Treatments

Hardware treatments are generally of permanent type provided for improvement of relief, physiographic and drainage features of watershed. These are erected with major government support with the purpose to check soil erosion, regulate overland flow and reduce peak flow. At times they are imposed to completely divert the upland runoff from running into downstream fertile lands. Different hardware treatments are discussed as follows:

(i) **Waterways:** Quite often, high runoff volumes are observed, even in drylands particularly in watershed located in sloppy areas. Such runoff water should therefore, be channelized through a few waterways. Some of the existing waterways may be developed as far as possible. Waterways draining larger areas may be designed on hydrologic and hydraulic considerations. No doubt, at places sizeable land strip may be occupied by such waterways but this may be found highly advantageous for the survival of trees planted along the waterways. Depending upon land situations, alignment of small size waterways may be adjusted with field boundaries. Such water courses may also be safeguarded against erosion by providing mechanical checks and also by raising vegetation which may be useful fodder for animals.

(ii) **Bunds:** These are low height earthen embankments constructed across the slope in cultivated lands after deciding location of waterways. The bunds function to intercept runoff, increase infiltration opportunity time and dispose excess rainfall safely. Such bunds can be either contour bunds or graded bunds, many a times these bunds are adjusted with field boundaries with field if deviation from grade or contour is not too much. Spacing between the bunds depends on many engineering considerations.

(iii) **Contour Bunds:** These are recommended in dry farming areas with light textured soils of slopes upto 6 per cent and where annual rainfall does not exceed 600 mm. They are designed for an expected runoff of 24 h duration and 10 yr frequency. Surplussing arrangements are provided to dispose off the excess runoff beyond the design storage. The cross section area of contour bunds follows the depth and type of soil, however, 0.5 m^2 is minimum.

(iv) **Graded Bunds:** These are constructed in medium to high rainfall areas in poor permeable soils having 2 to 6 per cent slope. They are also quite suitable for the soils having crust formation tendency. By and large graded bunds with 0.3 to 0.5 m^2 section are constructed with longitudinal gradient of 0.2 to 0.4 per cent depending on the site condition. Graded bunds particularly are made to reduce runoff, to reduce soil loss and to divert runoff.

For making bunds, required soil may generally be taken from lower side. Sometimes, excavations and pits for making bunds in deep black soil areas particularly are made on upper side. But these may be got levelled or ploughed up as early as possible after bunds construction. Invariably barring a few situations borrow pits should be made on lower side at the place where the bunds drain into waterways. In real field situations, bunds are covered with natural grass and in most cases, these are dependable sources of forage. In addition, quite often some trees also come up in their close vicinity, to the liking of cultivators. In some places loose stones are easily available, hence stone bunds are quite advisable in such pockets.

(v) **Terracing:** Cultivated lands having land slopes above 10 per cent particularly in hilly areas should be put under bench terracing by converting the lands into series of platforms. The width of bench terrace depends on the land slope and the permissible depth of cut. Bench terracing is very much effective in reducing soil erosion in hilly areas. At places where stones are available locally, loose stone walls can be made to act as risers for bench terrace construction.

(b) Medium Software Treatments

Medium softwares are also provided particularly as inter bunds where field sizes are large and conventional bunds are constructed along field boundaries. Such treatments are usually of semi permanent type and are adopted to minimize the velocity of overland flow. Such measures may last 2 to 10 yr. The important medium software treatments are discussed below:

(i) **Small Section Key Line Bunds:** A small section bund may be created across the slope at half of the vertical bund spacing. Such bunds can be nearly 0.1 m^2 in section; may be renovated at an interval of 2 to 3 yr.

(ii) **Strip Levelling:** About 4 to 5 m strip of land above the bund across the major land slope may be levelled for the purpose. Similarly, one or more strips can be created by running blade harrow after ploughing the field with mould board or disc plough. Such levelling programme may be taken up after every 2 to 4 yr.

(iii) **Live Beds:** One or two live beds (2–3 m wide) can be created either on contour or on grade in the inter bund space. The vegetation on the beds may be according to the liking of cultivator; it can be annual or perennial or combination of both.

(iv) **Vegetative Barriers:** One or two barriers of close growing crops or legumes can be created along the bunds as well as at the mid length of slope to filter the runoff water or slow down overland flow. *Khus* could be one of such vegetations. Several other promising grass species that serve as valuable fodder for cattle are being explored as an alternative to *khus* grass. A miniature bund at lower side of barrier is recommended to help in the development of live barriers, particularly in the initial stages.

(c) Software Treatment

Software treatments are taken up for ensuring fairly uniform soil moisture for satisfactory crop production. By and large, software treatments are temporary in nature; in that case these are required to be remade or renovated every year. The entire cost of applying such treatments is to be borne by the farmers. Because of their cost effectiveness, some of these have gained wide acceptance by the farmers in the recent years. The important software treatments are discussed as follows:

(i) **Contour Farming:** Contour farming is one of the easiest and most effective and low cost method of controlling erosion and conserving soil moisture. With contour farming, tillage operations are carried out along contours. It creates numerous ridges and furrows for harvesting sizeable amount of runoff inside the field itself.

(ii) **Compartmental Bunds:** Compartmental bunds, converting the area into square /rectangular plots are useful for temporary impounding of water for improving moisture status of the soil. These are made using bund formers. In medium deep black soil, they are found advantageous in storing the rainfall received during the rainy season in the soil profile, thereby augmenting the soil moisture for use by *rabi* crops. The size of the compartments may be fixed considering the size of the inter bunded land.

(iii) **Broad Bed and Furrows:** Broad bed and furrow system implies shaping alternate bed and furrow. This technique is especially suited to black soils, where crops are sown on pre formed beds. This system is made before the season and is maintained year after year. The planting is done on the beds. Generally, the depth of each furrow is kept 0.15 m and the inter furrow spacing is maintained at 1.5m.

(iv) **Dead Furrows:** Dead furrows are laid across the land slope in rolling lands to intercept the runoff. The spacing between dead furrows varies between 2 to 5 m or 4 to 7 crop rows. This system works well in alfisols.

(v) **Tillage:** Tillage operations help in entry of rain water in the soil. Off season tillage (summer ploughing), in particular, has been found quite useful in most of the rainfed areas. The tillage operations make the soil receptive to rainfall. This practice is very useful in light soils often prone to crusting.

(vi) **Mulching:** Surface mulching protect soil against beating action of rain drops and it also increases water infiltration into the soil. Further it helps in minimizing water evaporation from soil surface. Sometimes dry soil mulch (dust mulch) created simply by stirring the soil has been found very effective in conservation of soil moisture.

8.8.1.2 Soil and Water Conservation in Non–arable Lands

Control grazing is one of the simplest approach in reducing soil loss from denuded sloping lands. As seen from the experiences of Operational Research Project (ORP), soil and water conservation measures are required even in the non-agricultural/arable lands. These practices include contour and staggered trenches and contour furrows. They help in accelerating vegetation establishment, encouraging natural regeneration of species and combating soil erosion. Contour and staggered trenches are usually spaced 10 to 20 m apart across the slope for raising forest species. In general, 0.45 m wide trenches with 0.45 m depth are made at regular intervals. Sometimes in place of trenches, contour furrows are formed at a spacing of 2 to 10 m. In rocky areas, crescent bunds of loose boulders are constructed in place of trenches. All the above stated measures are useful in slowly regaining the lost forest cover and fertility of most degraded grasslands. Community efforts in the right direction are likely to make dent in regreening such areas.

8.8.1.3 Water Harvesting Measures

Since time immemorial, several kinds of water harvesting structures have been in practice in our country. Till recently, local experiences and precedences were utilized for designing these structures. In order to bring cost effectiveness, efforts are also necessary to harmonize the designing of water harvesting structures based on traditional wisdom and scientific contemplations. In fact, farm ponds, surface water tanks and percolation tanks need to be viewed as conservation structures in addition to sources for supplemental irrigation. In real watershed situations, ponds may be located in conjunction with water ways, some may be of a few hundred cubic metre size, whereas couple of structures may be of bigger size. In selected pockets, integrated catchment cum command approach should be followed. To ensure proper storage in light soils, some of the ponds can be lined with sealant materials. All large ponds have to be constructed at Government cost. All such structures are likely to recharge ground water reservoir. When located in proper eco-cultural hydrologic matrix, in conjunction with other factors, they also help in reducing load on ground water aquifer. It may be advisable to work out design details of all the required structures in the first phase of the project, no matter construction of a few of these may have to be deffered, may be planned to be taken up in drought years to provide employment to the rural masses. The objectives of water harvesting structures are:

- To store water for supplemental & life saving irrigation
- To act as silt detention structure
- To recharge the ground water
- To raise aquaculture species; and
- Recreation and allied agricultural uses

Thus in most of the situations, in addition to reviewing traditional water harvesting systems, other measures are also needed. Some of such typical water harvesting structures adopted in watersheds are given below:

(i) **Minor Irrigation Tanks:** Minor irrigation tanks are constructed across the major streams with canal system for irrigation purpose by constructing low earthen dams. A narrow gorge should be preferred for making the dam in order to keep the ratio of earthwork to storage as minimum. The height of the dam may vary from 5 to 15 m. The tanks are provided with well designed regular and emergency spill ways for safety against side cutting. In micro watersheds water harvesting *bundhies,* similar to small scale irrigation tanks are also recommended. By and large, the water harvesting *bundhies* are not intergrated with extensive canal system.

(ii) **Farm Ponds:** Farm ponds are typical water harvesting structures constructed by raising an embankment across the flow direction or by excavating a pit or a combination of both. Dugout ponds in light soil require lining of the sides as well as the bottom with suitable sealants. This is not the case with heavy black soil. Considering the benefit of the harvested water, different type of linings, viz, brick, concrete, HDPE, etc. may be used.

(iii) **Nala Bunds and Percolation Tanks:** Nala bunds and percolation tanks are located in the nalas having permeable formations with the primary objective of recharging the ground water. A strict regulation on the silt load entering the downstream reservoirs is an additional advantage of percolation tanks. The percolation tanks encourage the digging up of wells downstream of recharged area for irrigation purposes. The percolation tanks are provided with emergency spillways for safe disposal of flow during floods.

(iv) **Stop Dams:** Stop dams are permanent engineering structures constructed for raising the water level in the nala for the purpose of providing life saving irrigation during drought periods. These are located over flat nalas at narrow gorges carrying high discharge of long durations. They are created over stable foundation conditions where hard murrum or rocks are encountered. For the stop dam, a site with larger water storage capacity would be desirable.

8.8.2 Crop Management

The rainfed areas of our country contribute about 44 per cent of the total food grain production. In the watersheds there is scope of increasing the overall productivity of crops by adopting improved agronomical practices. For achieving sustainability in crop production in the watersheds the improved crop production technologies to be adopted includes selection of suitable crops & varieties, seeding techniques, nutrient management, weed management and plant protection measures. Improved crop management technologies for dryland is given in another chapter in this book in detail, however, a brief is given as under.

(i) **Crop and Cropping System:** Monocropping of erosion permitting crop accelerate soil and water loss year after year. Intercepting, erosion permitting and erosion resistant crops and their rotation have been found effective for soil and water conservation. As the pulse crops are effective for soil conservation due to their smothering effect, they should be sown in time to develop adequate canopy by the time of peak rate of runoff.

Row crop of tall growing crops such as sorghum, maize, pearlmillet, etc. are not effective in conserving soil as they expose a large portion of the soil and hence they are known as erosion permitting crops. Whereas close growing crops such as cowpea, greengram, blackgram, groundnut, etc. which protect soil are known as erosion resisting crops as they are very effective in reducing soil loss by minimizing the impact of raindrops and obstruction in runoff. In areas where rainfall ranges from 500 to 700 mm, intercropping facilitates the growing of cereal + legume or legume + legume in a system of different maturity patterns. In areas with annual rainfall more than 750 mm with soil water storage capacity of 150 mm or more, sequential cropping is possible. In addition to suitable crops and cropping systems, short duration varieties having deeper root system should be adopted to stabilize production in watersheds.

(ii) **Field preparation (Tillage operation):** Tillage is a well known soil and water conservation practice which makes soil surface more permeable to increase infiltration of rainwater into the soil, which in turn reduce runoff, soil and nutrient losses and enhance crop yields. Deep tillage is (25–30 cm) assists in opening up of hard soil layers and improves infiltration of rain water. Deep tillage promotes better root system development during low rainfall years, leading to more efficient use of water and nutrients available in sub-soil layers. Off season tillage or pre monsoon tillage also has a marked impact on weed control and rain water intake. Conservation tillage, which ensure at least 30 per cent converge of the soil surface with crop residue play very important role in organic carbon build up and soil and moisture conservation under drylands. Conservation tillage includes stubble mulch tillage, minimum tillage and zero tillage is very effective in soil and water conservation. Low intensity tillage favours consolidation of soils and imparts erosion resistance through better consolidation, structure, infiltration, pores distribution and depression storage.

(iii) **Contour Farming:** A line joining the points of equal height is called contour. All the cultural practices such as plouhging, sowing, inter-cultivation, etc. done across the slope on contours reduce soil and water loss. By ploughing and sowing across the slope, each ridge of plough furrow and each row of the crop act as obstruction to the runoff and provide more time for water to enter into the soil leading to reduced soil and water loss.

(iv) **Strip Cropping:** It is a system of growing of few rows of erosion resisting crops and erosion permitting crops in alternate strips on contour (across the slope) with the objective of breaking long slopes to prevent soil loss and runoff. Close growing erosion resisting crops reduce the transporting and eroding power of water by obstructing runoff and filtering sediment from runoff to retain in the field. Width of erosion permitting and erosion resisting crops vary as per the slope of the field. Strip cropping resembles the intercropping.

With increase in per cent slope of the soil, width of erosion permitting and erosion resisting crops decreases. The normal ratio between the erosion resisting crops and erosion permitting crops is 1:3. The strip cropping is simple, cheap and effective soil conservation practice and can be adopted by the farmers.

(v) **Mulching:** Mulching of soil with available plant residues reduce soil loss considerably by protecting the soil from direct impact of rain drops and reducing the sediment carried with runoff. A minimum plant residue cover of 30 per cent is necessary to keep runoff and soil loss within the acceptable limits. Vertical mulching also reduces soil loss particularly in vertisols by increasing infiltration.

(vi) **Cover Crops:** Good ground cover by canopy gives the protection to the land like an umbrella and minimizes soil erosion. Besides conserving soil and moisture, the cover crops hold those soluble nutrients which are lost by leaching, in the addition of organic matter build up. The important cover crops are cowpea, dhaincha, sunnhemp, greengram, blackgram, etc.

(vii) **Intergrated Nutrient Management:** Inadequate supply of plant nutrients and continuous use of inorganic fertilisers resulted in soil degradation. Soil health restoration should be the major factor in sustaining crop productivity in rainfed agriculture. Use of organic manures and bio-fertilisers along with limited use of inorganic fertilizers appears ideal for restoring the degraded soils for sustained productivity in rainfed areas.

8.8.3 Alternate Land Use Systems

It aims at land utilization according to its capability. About 10–30 per cent area in any of the watershed is found unfit for raising conventional crops due to soil degradation. There are two approaches in this direction. One is to put land belonging to Class IV and above under suitable grasses, fodder crops, fruit and forest species integrated with animal production system. Alternately, the same land or a portion of it could be simultaneously or sequentially put to food, fodder, and fuel or forest trees production to serve immediate benefits as well as to meet long term environmental concerns. Ley farming, alley farming, agri–horticulture, horti silvipastoral, tree farming, agri–silviculture are some of the alternate uses of lands.

(i) **Ley Farming:** It is the rotation of annual or perennial field crops with temporary pastures for the purpose of regenerating the fertility of the soil and improving the supply of forage. In this way ley farming is useful under Indian conditions where under dryland situation farmers are reluctant to apply fertilizers. Ley farming has several advantages such as pasture legume constitute high quality pasture feed, the pasture needs to be sown only once, favourable effects on soil protection and soil improvement and ultimately increase and stabilize the income of farmers. Studies at CRIDA, Hyderabad led to the identification of *Dolichos, Macroptilium, Leucaena, Clitora, Desmodium, Detrocema, Glycine, Calapogonium, Vigna, Tephrosia, Alysicarpus, Indigofera, Cassia and Stylosanthes.* Ideal legume among these is *Stylosanthes* which is high yielding, persistent, nutritive and palatable. *Stylosanthes hamata* is as remunerative as grain sorghum. *Stylosanthes hamata* and *Stylosanthes scabra* could also be grown as intercrops with dryland cereals such as sorghum, pearlmillet, fingermillet or along with range land grasses like *Dicanthium, Urochola, Lasiurus, Sehima, Cenchurus,* etc.

(ii) **Alley Cropping:** Alley cropping is essentially an agro-forestry system in which food crops are grown in alleys formed by hedge rows of trees or shrubs in arable lands. It is also known as hedge row intercropping. Hedge rows are cut back at about one meter height at planting and kept pruned during cropping to prevent shading and to reduce competition with food crops. It is recommended for humid tropics, primarily as an alternative to shifting cultivation. In semi-arid region of India, alley cropping provides fodder during dry periods since mulching the crop with hedgerow pruning usually does not contribute to increased crop production. There are three versions of alley cropping system which are forage alley cropping, forage cum-mulch alley cropping and forage cum sole alley cropping. Thus, there are two essential components, the hedge row and the crop. To optimize production, management of both becomes extremely important.

(iii) **Agri–Horticultural System:** It is one form of agro–forestry in which the tree component is fruit tree. It is also called as food cum–fruit system in which short duration arable crops are raised in the interspaces of fruit trees. Some of the fruit trees that can be considered are guava, pomegranate, aonla and ber. Pulses are the important arable crops for this system. However, depending on the requirements, others like sorghum and pearlmillet can be grown in the interspaces of fruit trees. The crops grown in the interspaces should be normally low statured and of short duration. This system not only keeps the interspaces weed free but also provides higher returns per unit area. In addition there could be improved soil fertility and availability of leafy fodder as well as fuel wood from the pruning.

(iv) **Agri-silviculture:** In this alternate land use system trees and agricultural crops grown together in same piece of land at the same time. Tree component gives fodder fuel or timber, including green leaf manure. It is ideal for class IV soils of drylands with annual rainfall around 750 mm. The positive associative effect of *Leucaena leucocephala* has been found in crops like hybrid napier, lucerne, oat and several other cereal forages.

(v) **Horti/silvi Pastoral System:** It is an agro-forestry system involving integration of fruit trees with pasture. When fruit tree is replaced by a top-feed tree, it is called as silvi-pastoral system. Guava, aonla and ber suits well in a horti-pastoral system with grasses like *Cenchrus ciliaris* (anjan), *C. setigerus* (bird wood), *Dicanthium annulatum* (marvel), *Chloris gayana* (rhodes) and *Panicum antidotale* (blue panic) and legumes like *Stylosanthes hamata,* S. scabra (stylo) and *Macroptilum atropurpuream* (stratro) etc. Under silvi-pastoral system the selected tree species should be a fast growing, multipurpose with leaves of high digestibility and palatability. It should also have a good competing ability trees ideal for silvi pastoral system are *Acacia nilotica (babul), Acacia senegal (gum acacia), Dalbergia sissoo* (rosewood or sheesham), *Sesbania grandiflora (agathi)* and *Leaucena leucocephala (subabul.)*

(vi) **Tree Farming:** Drylands not fit for cultivation (non-arable) and when fodder production is not a priority, can be put under tree farming as block plantation. The uncultivated wastelands which lack soil of any kind cannot support any vegetation. These could be utilized and improved by planting a space tree cover in selected microsites. These can flourish and yield abundantly in many places where arable crops would fail to grow. Tree farming resulting in other intangible benefits like water and soil conservation and moderating influence on the extreme variation in climate, common feature in many drylands.

Trees are also farm components of arable farming. They can be used as shelterbelts or for planting along the boundaries or on bunds along with arable crops. *Leucaena leucocephala (under short felling cycles), Prosopis cineraria, Zizyphus jujuba, albezia lebbeck* (higher felling cycles), etc. are suitable for successful tree farming.

(vii) **Silvi Pastoral System:** This system is suited to marginal drylands and is most preferable where the fodder shortage is more. This system essentially consists of a top feed tree species and grass or legumes (preferably perennial) as understory crops. Dryland farmers having larger holdings and keeping a part fallow for a longer period for one reason or the other should go in for this system which could provide both fodder and fuel. In survey carried out in Andhra Pradesh, Karnataka and Maharashtra by CRIDA scientists, it was revealed that after food, it is the fodder that is of paramount importance for sustaining animal wealth in rural areas. Appropriate assessed grasses and legumes for drylands are given in table 8.2 as under.

Table 8.2: Suitable Grasses and Legumes for Silvi-pastoral Systems

Grasses	Legumes
Cenchrus dhotis (buffel grass), *Cenchrus setigerus* (bird wood grass), *Dicanthium annulatum,* (marvel) *Chloris gayana* (rhodes grass) *sorghum sudanese* (sudan grass) *Panicum antidotale,* (bluepanic grass) *Panicum coloratum* (klein grass or blue panic grass), *Panicum maxima* (guinea grass), *Pennisetun pedicellatum* (dinanath grass.)	*Stylosanthes hamata* (stylo), *Styolsanthes sacbra* (stylo), *Dolichos lablab* (fieldbean), *Macroptillium atropurpureum* (siratro), *Cyamopsis tetragonoloba* (clusterbean,) *Vigna unguiculata* (cowpea), *Dolichos biflorus* (horsegram).

A brief about different land use options/ farming systems according to annual rainfall and soil type is given is table 8.3 as under :

Table 8.3: Alternate Land Use Options for Different Agro-climatic Situations.

Annual Rainfall (mm)	Soil type	Farming systems
Less than 500	Shallow (0–30 cm)	**Tree farming** *Prosophis cineraria, Prosophis juliflora, Acacia tortilis, Pithecellobium dulce*
	Medium (0–45 cm)	**Pasture management** *Lasiurus sindicus* (light textured soils), *Cenchrus setigerus, Sehima nervosum, Sylosanthes scabra,* etc.
500–750	Shallow (0–30 cm)	**Silvi pastoral system** *Acacia nilotica, Dalbergia sissoo, Hardwickia binata, Cassia sturti, Albezia amara, Leucaena* spp., *Cenchrus ciliaris, Cenchrus setigerus, Dicanthimum annulatum, Panicum antidotale, Stylosanthes hamata,* etc.
	Medium (0–45 cm)	**Horti-pastoral system** *Custard apple, ber, jamun, anola, tamarind, wood apple, bael, Cenchrus ciliaris, Panicum antidotale, Stylosanthes hamata, , Clitoria ternata*

Annual Rainfall (mm)	Soil type	Farming systems
More than 750	Shallow (0–30 cm)	**Ley farming or Silvi–pastoral system** 3 years Stylosanthes hamata and 4th year arable crops (sorghum on heavier soils; pearlmillet on lighter soils), Silvi–pastoral system as above.
	Medium (0–45 cm)	**Ley farming or Horti–pastoral system.** Ley farming, mango, sapota, guava, aonla, *Stylosanthes hamata, Macroptilium atropurpureum.*

8.9 Factors Affecting Watershed Management

8.9.1 Watershed Characteristics

Certain physical properties of watersheds significantly affect the characteristics of runoff and as such are of great interest in hydrologic analysis. Watershed characteristics *i.e.* size, shape, topography, slope, vegetative cover and drainage density influence the watershed management as follows:

(i) **Size and Shape:** Size of watershed determines the quantity of rainfall received, retained and surface runoff. A small watershed is pronounced by overland flow which is the main contributor to peak flow. A large watershed has no overland flow, but channel flow is significant. Large watersheds are also affected by basin storage.

Shape contributes to the speed with which the runoff reaches the river. A long catchment area will take longer to drain than circular catchment. Basin shape is not usually used directly in hydrologic design methods. Watersheds have infinite variety of shapes and the shape supposedly reflects the way that runoff will bunch up at the outlet. A circular watershed would result in runoff from various parts of the watershed reaching the outlet at the same time. An elliptical watershed having the outlet at one end of the major axis and having the same area as the circular watershed would cause the runoff to spread out over time, thus producing a smaller flood peak than that of circular watershed.

(ii) **Topography:** Topographic configuration such as slope, length, degree and uniformity of slope affect both disposal of water and soil loss. Time of concentration and infiltration of water are thus a function of degree and length of slope of the watershed. Topography determines the speed with which the runoff will reach a river. Clearly, rain that falls in steep mountainous areas will reach the primary river in the watershed faster than flat or slightly sloping areas.

(iii) **Vegetative cover:** Vegetative cover is an important landscape element in any watershed. Distribution of vegetation species may be diverse and highly variable across the watershed, but vegetation communities can be described in more general terms as well.

Depending upon the type of vegetation and its extent, this factor regulates the functioning of watershed (infiltration, water retention, runoff production, erosion sedimentation, etc). Vegetation plays vital role in the water cycle. It intercepts rainfall, impedes overland flow and promotes infiltration. Vegetation also uses water for growth. All of these factors reduce the quantity of runoff to streams. Vegetation binds and stabilizes soil, thereby reducing the soil erosion. Vegetation functions to slow runoff and reduce soil compaction, allowing better percolation of rainfall into soil (infiltration) and groundwater recharge, which creates better water storage for summer base flows. Root systems also help to keep soil material stable from moving down slope, provides protection to the soil layer.

(iv) **Drainage Density:** Drainage density is the total length of all the streams and rivers in a drainage basin divided by the total area of the drainage basin. It is a measure of how well or how poorly a watershed is drained by stream channels. It is equal to the reciprocal of the constant of channel maintenance and equal to the reciprocal of two times the length of overland flow.

Drainage density can affect the shape of a river hydrograph during a rainstorm. Ricers that have a high drainage density have a flashier with a steep falling limb. High densities can also indicate a greater flood risk.

8.9.2 Climatic Characteristics

The greatest factor controlling stream flow, by far, is the amount of precipitation that falls in the watershed is rain or snow. However, not all precipitation that falls in a watershed flows out and a stream will often continue to flow where there is no direct runoff from recent precipitation.

Climatic parameters affect watershed functioning and its manipulation in two ways. Rain provides incoming precipitation temporarily and spatially along with its various characteristics like intensity and frequency. The amount of rainfall and these parameters along with temperature, humidity, wind velocity, etc. regulates factors like soil and vegetation. Soil properties reflect the climate of the region. In the same way, the vegetation type of a region depends totally on the type of the climate.

8.9.3 Land Use Pattern

Type of land use, its extent and management are the key factors which affect watershed behaviour. Judicious land use by users is of vital importance to watershed management and functioning. Change of land use within the watershed, especially within the variable source area, greatly affects the collection capacity and consequent runoff behaviour of the watershed. The extent of land use change over the watershed has effects that are similar to the relationship between areal storm extent and watershed size. If the land use changes are local, then the impact of such changes is especially apparent in the storm hydrograph. The storm hydrograph is dominated by local characteristics. For land use changes that cover larger portion of the watershed, the impacts may also be observed in the annual hydrograph.

Watershed maintain the health of forests, fisheries, wetlands, coastal resources, agricultural landscape, habitat and local communities. Protecting watersheds ensure a sustainable environment that supports recreational activities and a healthy economy.

8.9.4 Social Status

Common contributors to water pollution are nutrients and sediments which typically enter in the stream systems after rainfall, washes them off poorly managed agricultural fields called surface runoff or flushes them out of the soil through leaching. These types of pollutants are considered non-point sources of pollution because the exact point where the pollutants originated cannot be identified. Such pollutants remain a major issue for water ways because of the difficulty to control their sources, hinders any attempt to limit the pollution. Point source pollution originates at a specific point of contamination such as if a manure containment structure fails and its contents enter the drainage system.

8.10 Major Watershed Development Projects in India

A brief account of various soil and water conservation programmes/ projects executed by Central/ State Governments are discussed in brief as under:

8.10.1 Drought Prone Area Programme (DPAP)

Drought Prone Area Programme (DPAP) was started in 1973 as an intergrated area development programme. This programme was considered as a long term measure for restoration of ecological balance and optimum utilization of water, livestock and human resources to mitigate the ill effects of drought. The aims of the programme are (i) to minimize the adverse effects of drought on production of crops and livestock and productivity of land, water and human resource base of the area and adoption of appropriate technologies, and (ii) to conserve, develop and harness land, water and other natural resources including rainfall for restoration of ecological balance in the long run. The Drought Prone Area Programme operated in 615 blocks of 95 districts in 13 states. The total area covered by the scheme was 53.6 Mha.

8.10.2 Desert Development Programme (DDP)

This Desert Development Programme was initiated in 1977-78 with the objectives of (i) controlling desertification of the desert area through integrating and dovetailing other related State/ Central programmes and (ii) to conserve, develop and harness land, water and other natural resources including rainfall for restoration of ecological balance. The aims of the programme are sought to be achieved through activities like afforestation, with special emphasis on sand dune stabilization, shelterbelt plantation and grassland development, soil and moisture conservation and water resources development. This programme covers 131 blocks of 21 districts in five states namely Rajasthan, Haryana, Gujarat (hot arid region), Jammu and Kashmir and Himachal Pradesh (cold arid region). The total area covered under the Desert Development Programme was about 36.2 Mha. This project was implemented with 100 per cent Central finance assistance.

8.10.3 National Watershed Development Project for Rainfed Areas (NWDPRA)

The National Watershed Development Project for Rainfed Areas was initiated in the year 1986–87. This programme was implemented in the unirrgated arable lands mostly falling in the rainfall range of 500 to 1125 mm and also above. The programme covers 25 states. The districts having more than 30 per cent area under irrigation are usually excluded.

In 8th Five Year Plan the project was restructured to achieve the twin objectives of sustainable production of bio-mass and restoration of ecological balance in vast rainfed areas of the country.

It has main focus on (i) conservation, upgradation and utilization of natural endowments such as land, water, plant, animal and human resources in a harmonious and intergrated manner with low cost, simple, effective and replicable technology (ii) generation of massive employment (iii) reduction of inequalities between irrigated and rainfed areas (iv) improvement of production environment and restoration of ecological balance through scientific management of land and rainwater endeavour to enhance cash flow to the rainfed farmers and landless agricultural labourers through increased casual employment, marketable surplus of agricultural and dairy produce and growing of cash crops in suitable areas. Thus the ultimate objective of this project was to develop the natural resources base, sustain its productivity, improve the standard of living of millions of poor farmers and landless labourers and endeavour for restoration of ecological balance.

The main components of the project were (i) land and moisture management including scientifically tuned cropping systems, dryland horticulture, fodder production and farm forestry (ii) contingency seed and planting material stocking (iii) training, seminars, study tours for staff and farmers within the state/ regional/ national level (iv) adoptive research trials on different crops in small and marginal farmers land (v) procurement, fabrication and supply of survey equipments and prototype implements, and (vi) preparation of field manuals and publicity materials.

The Intergrated Watershed Development Plans of the identified areas were developed and executed through people's participation with the help of *Mitra Kisans* and *Gopals*. The project is being implemented by the participating state governments through 100 per cent Central assistance comprising 75 per cent grant in aid and 25 per cent loan to the states.

8.10.4 Integrated Watershed Management Programme (IWMP)

The Department of Land Resources, Ministry of Rural Development, Government of India implements Centrally sponsored programmes of Intergrated Wasteland Development Programme (IWDP), Drought Prone Areas Programme

(DPAP) and Desert Development Programme (DDP) for development of waste/ degraded lands through watershed approach. The three programmes were implemented during 2009-10 on the basis of separate norms, funding patterns and technical components based on their respective objectives. While DDP focused on reforestation to arrest the growth of hot and cold deserts, DPAP concentrated on agroforestry, pasture development, horticulture and alternate land uses. IWDP, on the other hand, adopted pasture development, soil and moisture conservation as prominent activities on wasteland under government, community or private control. The common theme was sustainable management of land and water resources.

The main objectives of the IWMP are to restore the ecological balance by harnessing, conserving and developing degraded natural resources such as soil, vegetative cover and water. The outcomes are prevention of soil runoff, regeneration of natural vegetation, rain water harvesting and recharging of the ground water table. This enables multicropping and the introduction of diverse agro based activities, which help to provide sustainable livelihoods to the people residing in the watershed area.

In order to bring about integration of the entire area development, a new programme with the title Intergrated Watershed Management Programme has been launched for intergrated planning, sustainable outcomes, and rural livelihoods of the communities. All the three area development programmes has been covered under Intergrated Watershed Management Programme which is to be implemented by dedicated agencies, which will be operational at national, state and district levels.

8.10.5 Watershed Development Component of Prime Minister Krishi Sinchayee Yojna (WDC-PMKSY) erstwhile (IWMP)

For optimum use of resources, sustainable outcomes and integrated planning, DPAP, DDP and IWDP were consolidated as the Watershed Development Component of Prime Minister Krishi Sinchayee Yojna (WDC-PMKSY) was initiated during the year 2015.

The main objectives of the WDC-PMKSY are to restore the ecological balance by harnessing, conserving and developing degraded natural resources such as soil, vegetative cover and water. The outcomes are prevention of soil erosion, regeneration of natural vegetation, rain water harvesting and recharging of the ground water table. This enables multi-cropping and the introduction of diverse agro-based activities, which help to provide sustainable livelihoods to the people residing in the watershed area.

The Salient features of WDC-PMKSY are as follows:

- Setting up of dedicated institutions with multi-disciplinary experts at State level - State Level Nodal Agency (SLNA), District level - Watershed Cell cum Data Centre (WCDC), Project level - Project Implementing Agency (PIA) and Village level - Watershed Committee (WC).

- Cluster approach in selection and preparation of projects: Average size of project is about 5,000 ha.
- Enhanced cost norms from Rs. 6000 per ha to Rs.12,000 per ha in plains; Rs. 15,000 per ha in difficult/hilly areas.
- Uniform funding pattern of 90:10 between Centre and States.
- Release of central assistance in three installments (20%, 50% & 30%) instead of five installments.
- Flexibility in the project period *i.e.* 4 to 7 yr.
- Scientific planning of the projects by using IT, remote sensing techniques, GIS facilities for planning and monitoring and evaluation.
- Earmarking of project funds for DPR preparation (1%), entry point activities (4%), capacity building (5%), monitoring (1%) and evaluation (1%).
- Introduction of new livelihood component with earmarking of project fund under Watershed Projects *i.e.* 9 per cent of project fund for livelihoods for assetless people and 10 per cent for production system and micro-enterprises.
- Delegation of power of sanction of projects to states.

8.10.6 Neeranchal Watershed Program

Neeranchal is a World Bank assisted National Watershed Management Project. Neeranchal is designed to further strengthen and provide technical assistance to the Watershed Component of PMKSY, in particular and all components of PMKSY, in general, to enhance its delivery capacity. The programme is being implemented in nine participating states - Andhra Pradesh, Chhattisgarh, Gujarat, Jharkhand, Madhya Pradesh, Maharashtra, Odisha, Rajasthan and Telangana during the year 2015.

For achieving the major objectives of the Watershed Component of the Pradhan Mantri Krishi Sinchayee Yojana (PMKSY) and for ensuring access to irrigation to every farm (*Har Khet Ko Pani*) and efficient use of water (per drop more crop). Neeranchal is primarily designed to address the following concerns:

- To bring about institutional changes in watershed and rainfed agricultural management practices in India.
- To build systems that ensure watershed programmes and rainfed irrigation management practices are better focused, and more coordinated, and have quantifiable results.
- To devise strategies for the sustainability of improved watershed management practices in programme areas, even after the withdrawal of project support through the watershed plus approach, support improved equity, livelihoods, and incomes through forward linkages, on a platform of inclusiveness and local participation.

◆

GLOSSARY

Abiotic factors: Physical, chemical and other non living components of the environment such as light, temperature, moisture, nutrients and other edaphic factors.

Absolute drought: A period of at least fifteen consecutive days none of which has received as much as 0.25 milli meters of rain.

Absolute humidity: It is the amount of water vapour present in a unit volume of air.

Afforestation: Artificial establishment of a forest on land, which had not been previously grown tree crops or on deforested land.

Agricultural drought: It occurs where soil moisture and rainfall are inadequate during the growing season to support healthy crop growth to maturity and cause extreme crop stress and wilt.

Agri-silviculture: A combination of forestry and agricultural crop production either simultaneously or in rotation.

Agri-silvi-pastoral-system: Growing of trees, agriculture crops and grasses together in same land the same time. It improves and sustains the crop productivity which increases the level of income of the farmers.

Agro-climatic regions: The grouping of different physical areas within the country into broadly homogenous zones based on climatic and edaphic factors.

Agro-forestry: A collective term for systems of land use in which woody plants (trees and shrubs) are deliberately combined on the same land management unit with herbaceous crops and or animals; either in some form of spatial arrangement or in sequence. For a land use system to fall within the concept of agro-forestry, there should be both ecological and economic interactions between woody plants and other components of the system. It may involve the integration of trees into farming systems of crops and livestock into forests.

Antitranspirants: Substance that retard or inhibit transpiration. Examples are PMA, Cycocel, Kaolin, etc.

Arable crop: Annual crops which require the land to be ploughed and cultivated (unless direct drilled) and seeds to be sown as distinct from those which yield a product without such annual activities (*i.e.* fruit, grass, etc.)

Arable land: Land capable of being ploughed that is fit for tillage as opposed to that only suited for pasture.

Arid: Applied to a climate or a region where the rainfall is barely sufficient to support vegetation; deficient in rainfall.

Arid climate: A generally extremely dry climate with an annual mean precapitation usually less than 250 mm and generally rainfall is well short of evapo-transpiration demand of the atmosphere.

Arid zone: A zone of very low rainfall with most of the deserts.

Aridity index: (i) A measure of dryness of a region and is expressed as; Al = (no of rainy days x mean precipitation day) / (mean temperature + 0), (ii) The ratio of the water deficit to the water need (D/PE) usually expressed as percentage.

Aridity: (i) Area having less than 250 mm of rains per annum (ii) area where the total evaporation potential exceeds the actual precipitation (iii) Inadequate rainfall to support agriculture without irrigation.

Atmospheric moisture: Water as it occurs in various forms in the atmosphere.

Available soil moisture: The amount of water in the soil at any time in excess of the wilting coefficient, expressed either as percentage by weight of dry soil or as equivalent of available water.

Available water holding capacity: The capacity of a soil to hold water available for plant use. It can be calculated by the formula: AWHC= Pvd/100. Where P= per cent moisture of the soil held between field capacity and wilting point; v= volume weight of the soil expressed in grams per cubic centimeters; and d= depth of soil in centimeters. AWHC is expressed in centimeters.

Available water: The portion of water in a soil that can be readily absorbed by plant roots. Considered by most workers to be that held in the soil with a water potential of 1/3 up to approximately 15 bars.

Basin recharge: That portion of the storm rainfall that is required to satisfy the demands of interception depression storage and soil moisture. It is the difference between rainfall and runoff and is often referred to as loss.

Bench terrace: An embankment constructed across slopping fields with a steep drop on the down slope side.

Border strip cropping: A kind of strip cropping involving growing of strips of permanent vegetation placed around the border of a field, a farm or body of water. The border strip may be used to control either wind or water erosion. It is especially useful to filter the silt out of runoff water before it reaches between feedlots and streams.

Border strip: A grassed thickly vegetated strip at the edge of a field, along outlet channels or at ends of rows to check or prevent erosion.

Broad bed and furrow (BBF) system: A land management system for successful crop production in vertisols (heavy black soil) by providing better drainage through rainfed broad beds (about 90–100 cm wide) and furrow (25–30 cm wide)

Broad base terrace: A ridge type terrace 25–50 cm high and 5 to 10 meters wide with gently sloping sides, a rounded crown and dish shaped channel along the upper side, constructed to control erosion by diverting run off along the contour at a non scouring velocity.

Buffer strips: Contour strips of grass or other erosion resisting vegetation or below cultivated strips or fields.

Bund: An artificial earthen embankment made across sloping agricultural land to cut short lengthy soil slopes and reduce runoff and erosion.

CAM plants: The CAM (Crassulacean Acid Metabolism) system is prevalent in desert plants where the Co_2 is fixed at night in these species, there is diurnal fluctuation of acidity in thickened leaves. The CAM plants are adopted to environment or more or less constant aridity, e.g. *Bryophyllum* Spp. and cactus plants.

Capillary moisture: Water held by surface tension in the capillary spaces and as a continuous film around the particles, free to move under the influence of capillary forces, and available to plants. It is held with a tension of 60 cm of water. It moves from thick to thin films and constitutes soil solution.

Capillary movement: Movement of soil moisture in any direction caused by capillary tension.

Check dam: A small low dam constructed in a gully or other water course to decrease the velocity of stream flow and thereby to minimize channel scour and promote deposition of eroded material.

Cloud seeding: Injecting the clouds with a seeding agent like dry ice, silver iodide, sodium chloride from an air craft or using a ground generator for producing artificial rain.

Command area: The irrigable land area by the canal or from a dam of an irrigation project.

Community forestry: It is the forest of the community by the community and for the community. Such foresty is managed for the interest of local community or a village and is run almost exclusively by them.

Conservation farming: The most efficient use of the land over a long period of time which protects and preserves the land for agriculture.

Conservation tillage: Tillage designed to maintain roughness of a field surface while providing a suitable seed bed and weed control for the next crop. The roughness reduce water runoff and water and wind erosion.

Contingent cropping: Contingent cropping is growing of crops in aberrant situations like drought /floods. It aims at partial mitigation of misery by producing some food, feed and fodder to encounter emergency conditions.

Contour: An imaginary line connecting points of equal elevation of the surface of the soil. A contour terrace is laid out on a sloping soil at right angles to the direction of the slope and nearly level throughout its source.

Contour farming: A method of cultivation wherein operations including sowing are carried out along the contour. It reduces run off, conserves more moisture and increases crop yield.

Contour bunding: The construction of small bund across the slope of the land on a contour, so that the long slope is cut into a series of small ones and each contour bund acts as a barrier to the flow of water.

Contour interval: The differences in elevation between adjacent contours on a map.

Contour ploughing: Ploughing done along the contours.

Contour strip cropping: The cultivation practices involving growing of a soil exposing and erosion permitting crop in strips of suitable widths alternating with strips of soil protecting and erosion resisting crops along the contours.

Cover crops: A close growing crop grown primarily for the purpose of protecting and improving soil between periods of regular crop production or between trees and vines in orchards and vineyards. Cover crops are known to reduce the danger of soil erosion and to protect the land from leaching losses.

Crop equivalent yield: It is the conversion of crop yields into one form to compare the crops grown mixed /intercropped/ sequentially cropped. Conversion is done into monetary value/ protein /carbohydrate/ intrinsic energy value.

Crust: A surface layer on soil ranging in thickness from a few millimeters to perhaps as much as 3 cm that is much more compact, hard and brittle when dry than the material immediately beneath it.

Dead furrow: An open trench left in between adjacent strips of land after completion of ploughing.

Deep percolation: (i) The moisture which penetrates below the depths from which it may not be used by plants; it represents that part of the water absorbed which exceeds the field capacity of the soil within the depth of root development (ii) in a geologic science the amount of water that is lost from a basin by leakage through the geologic formation.

Deep tillage/ sub-soiling: It refers to tillage that loosens the soil 40 cm deep or more with the purpose of conserving moisture in the monsoon season or for breaking hard pans.

Desert soil: It often contains high percentage of soluble salts, high pH, poor in organic matter, water is scarce and in this area usually coarse millets are grown.

Drought avoidance: Permits a crop to grow for a longer time in a given moisture stress conditions (may be due to deep root system/uses less water).

Drought escaping varieties: Varieties which are relatively photo-insensitive and which have a shorter duration and the resulting ability to escape droughts. Early seeding vigour and good population performance are the other attributes of such strains.

Drought resistance: A characteristic of plants suitable for cultivation in dry conditions, no matter what mechanism they possess which provides resistance. One of the more important properties is their capacity to endure, without injury, an intense loss of water.

Drought: The deficit which results when the amount of water available in the soil is not sufficient to meet the demands of potential evapotranspiration. It is thus a condition where water need is in excess of available moisture.

Drought proofing: Conceptually, drought means the capacity to meet the basic material and physical needs of period so that there is minimal distress.

Drought year: A year in which rainfall of particular place is short by more than twice the standard deviation.

Dry land crops: Crops grown in dry farming areas such as barley, bajra, millet, pigeonpea, blackgram, mothbean, cowpea, clusterbean, sesamum, linseed, flax, castor, safflower, mustard, toria, taramira.

Dryland farming: (i) The practice of crop production in rainfall areas by conserving natural precipitation. Supplemental irrigation may be given from harvested runoff water. (ii) The practices of crop production without irrigation where rainfall is usually deficient.

Dryland: (i) Used in India to mean the opposite of wet lands, the latter lands are usually used for rice and are flooded (ii) used to describe land tracts which are primarily rainfed as opposed to irrigated.

Dry spell: A period of at least 15 consecutive days, during which not a single wetting rain has fallen.

Drainage: The removal of excess surface or ground water from land by means of surface or sub surface drains.

Dust mulch: An induced loose, fine, granular or powdery conditions in the surface soil; formerly believed to be effective in reducing evaporation losses of water from soil.

Effective rainfall: The proportion of total rainfall which is useful or utilizable for productive purposes. According to FAO it is that part of the total annual or seasonal rainfall which is useful directly and/or indirectly for crop production at the site where it falls, but without pumping. It, therefore, includes water intercepted by living or dry vegetation that is lost by evaporation from the soil surface, the precipitation lost by evapotranspiration and fraction which contributes to leaching and facilitates other cultural operations, either before or after sowing, without any harm to yield and quality of the principal crop.

Effective rooting depth: Soil depth from which the fully grown crop extracts most of the water need for evapo-transpiration.

Erodibilty: The lack of resistance of soil to soil erosion.

Erosion resistant crop: A crop, because of its dence foliage, root system, etc. provides effective protection against soil loss by erosion.

Erosion: The removal of soil from the land surface by water, wind or other forces.

Erosive: The ability of rainfall to cause soil erosion.

Evaporation: It is the total amount of water lost due to transpiration by a crop and evaporation from the soil surface during a specified time from a particular area.

Field capacity: The moisture percentage, on a dry weight basis of a soil after rapid drainage has taken place following an application reach of the root zone. This moisture percentage usually is reached within one to three days after an irrigation. The time interval depending on the physical characteristics of the soil.

Farm forestry: Farm forestry is a process in which trees are grown specifically for fuel, food and for a variety of other resources on a farm. It is a three dimensional farming and provides an alternative to monoculture grain farming. Three dimensions are defined as the trees themselves, the harvest from the trees which is used to feed livestock and the animals and their products.

Farm pond: A small body of water retained behind a small dam or held in a pit dug in the ground by water harvesting techniques.

Forest: A plant community predominantly of trees and other woody vegetation usually with close canopy.

Forestry: The act, occupation or art of farming and cultivating forests and systematic utilization, reproduction and improvement of the productive capacity of trees in masses, including the planting and culture of new forests.

Graded terrace: A terrace having a constant or variable grade along its length.

Grassed water way: A natural or constructed water way, usually broad and covered with erosion resistant grasses used to conduct surface water from crop land.

Grassland: The land on which graminaceous species represent the dominant if not the exclusive vegetation. Grassland is intermediate in status between forest or woodland on one hand and the desert on the other.

Hardening: Increase in the tolerance of a plant tissue to stress or adverse conditions.

Hectare metre water: Water column having vertical depth of one meter stored over one hectare area subject to no leaching, percolation, seepage or evaporation of water. It is equivalent to 10,000 cubic metre water, or 10,000 tonnes of water.

Humidity: A general term indicating the amount of water vapour in the air.

Ideotype: Refers to plant type in which morphological and physiological characteristics are ideally suited to achieve high production potential and yield reliability.

Incipient wilting: Small loss in turgidity during warm weather, even when soil is moist. It may not lead to any visible stress symptoms like rolling or folding of leaves and is usually reversible at night.

Intercropping: Refers to growing of two or more generally dissimilar crops simultaneously on the same piece of land, base crop necessarily in distinct row arrangement. The recommended optimum plant population of the base crop is suitably combined with an appropriate additional plant density of the associated crop and there is crop intensification in both time and space dimensions.

Interculture: Operations to be carried out after sowing or planting in between crop rows. Includes operations like gap filling, thinning, weeding, mulching and top dressing. May be done by manual labour or by bullock or machine drawn implements.

Infiltration: It is process of water entry into the soil generally (but not necessarily) through the soil surface and vertically downward.

***In situ*:** In the natural position or original position or location.

Intercrops: The crop raised in an orchard or other widely spaced crops for increasing the income from the same piece of land.

Jhum cultivation (Jhuming): The slash and burn type of shifting cultivation followed in the hill tracts of Bangladesh and Assam.

Kharif crops: Crop grown during main monsoon season (April to September) such as sorghum, maize, pearlmillet, rice, cotton, etc.

Land equivalent Ratio (LER): It denotes the relative land area under sole crop required to produce the same yield as obtained under a mixed or an intercropping system at the same management level. It is calculated as sum total of the ratios of yield of each component crop in an intercropping or a mixed cropping system to its corresponding yield when grown as a sole crop thus: LER = Yab/Yaa + Yba/Ybb. Where, Yab is the yield of Crop "a" in association with crop "b" and Yba is the yield of Crop "b" in association with Crop a, Yaa and Ybb represent the pure stand yield of Crop "a" and "b" respectively.

Ley farming: System of growing grass in rotation with agriculture crops; extensively used in U.K. and U.S.A.; helps to improve soil fertility and providers good fodder to animals; also helps in buildup of soil structure and prevent erosion. Can be conveniently adopted in places like Nilgiris abounding in good pasture land and having suitable climate.

Land capability classification: A grouping of the mapping units of soil conservation survey into land capability units, sub classes, classes and general divisions.

Life saving irrigation (Supplemental irrigation): The irrigation given to a dry land or rainfed crop to save it from total failure due to long dry spells or severe soil moisture stress, the source of irrigation generally being harvested water.

Lopping: Pruning or cutting away of aerial parts of shrubs or trees.

Minimum tillage: The soil manipulation necessary to meet the minimum tillage requirements for crop production.

Minor watershed: Part of watershed which may be selected as a suitable area for detailed planning and application of soil and water conservation measures in cooperation with relatively small local groups of land owners and operator, or local agencies as drainage districts of municipalities.

Mixed farming: Farming system which involves the raising of crop and rearing of animals and / or poultry. Mixed farming is based upon the principle that land should support animal and vice versa.

Mid season correction: It is a contingent management practice to overcome unexpected and or unfavourable weather conditions e.g. thinning, or population adjustment according to the availability of moisture in the soil during the dry spell.

Mixed cropping: Growing two or more crops simultaneously on the same piece of land without any definite row arrangement.

Moisture Availability Index: It is the ratio of assured rainfall (weekly or monthly) at 50 per cent probability level to potential evapo-transpiration of the corresponding period.

Moisture deficit index (MDI): This is the measure to estimate dryness of a region and is expressed by the formula: MDI + (P–PE) /PE) × 100 where MDI = Moisture deficit index, P= rainfall (mm), PE= Potential evapotranspiration (mm).

Moisture index: (i) The ratio of the plasticity index to the difference between the natural water content and the plastic limit of a soil (ii)The difference between precipitation (P) and potential evapo-transpiration (PE) divided by potential evapo-transpiration usually expressed in a percentage.

Moisture percentage: (i) Dry weight basis; the weight of water per 100 units of weight of material dried to constant weight at a standard temperature (ii) depth basis; the equivalent depth of free water per 100 units of depth of soil. Numerically this value approximates the volume of water per 100 units of volume of soil.

Moisture content (soil): The quantity of water present in soil usually expressed in percentage by weight, on oven dry basis.

Moisture retentive soil: A soil that retains a high percentage of water, directly depending upon clay soils are more moisture retentive than sandy soils.

Moisture stress: The tension at which water is held by the soil.

Monsoon: The seasonal wind of the Indian Ocean and Southern Asia, blowing from the southwest in summer and from the northeast in winter. Monsoon blows, commonly marked by heavy rains.

Mulch: (i) A layer of organic or inorganic material spread on the soil surface around plants to conserve moisture and suppress weeds. Organic mulches also add nutrients to the soil (ii) Any material such as straw, plant residues, leaves loose soil or plastic film placed on the soil surface to reduce evaporation, erosion or to protect plant roots from extremely low or high temperature.

Mulch farming: A system of farming in which organic residues or other materials are neither ploughed into the soil nor mixed with it, but are left on the surface to serve as mulch.

Mulch tillage: A crop and soil management practice that utilizes the residual mulches of the preceding crop by leaving a percentage of this crop residue on or near surface of the ground as protective mulch. Also called stubble mulching, mulch farming.

Non tillage planting: Direct planting in essentially unprepared seed beds.

Permanent pasture: A pasture of perennial or self seeding annual plants kept for grazing indefinitely.

Percolation: It is a down ward movement of water through the interstices of rock or soil under gravity or hydrodynamic pressure or both under saturated or nearly saturated condition.

Physiological drought: Non availability of water to the plants due to unfavorable physiological conditions, such as water logging, soil salinity, low temperature, etc.

Plastic mulch: Thin polythylene film, which may be transparent or black that is used as mulch for field crops and helps in moisture retention, increasing soil temperature and weed control.

Permanent wilting: A wilting condition of the plant due to loss of water. The plants would not recover from wilted condition during the night and they require watering for their recovery from wilted condition and regain of turgor.

Potential evapotranspiration: (i) The amount of moisture that will be evaporated from the soil and transpired from vegetation if it is available. It combines moisture and heat factor in climate. (ii) The evapotraspiration that occurs when the ground is completely covered by actively growing vegetation and there is no limitation of the soil moisture.

Precipitation: Total amount of precipitation (rain, drizzle, snow, hail, fog, condensation, hoar frost and rime) expressed in depth of water which would cover a horizontal plane if there is no runoff, infiltration or evapo-transpiration; mm/day.

Preparatory tillage: Various tillage operations carried out on the farm before the crop are sown, planted or transplanted to make a fine seedbed.

Rainy day: A period of 24 hours in which 2.5 mm or more rain is recorded.

Rainfed farming: Growing of field crop entirely with rain water received during the crop season under humid and sub humid climates and the crops may face little or no moisture stress during their life cycle.

Rainy day: A recorded rainfall of 2.5 mm or more within a period of 24 hours.

Rainy season: That period of the year in which monthly rainfall exceeds 10 per cent of the monthly potential evapotranspiration (in India May-June to September-October).

Relative humidity (RH): A measure of moisture content of air expressed as percentage. It is the ratio of the mass of water vapour in a given volume of air at a given temperature to the maximum quantity of water vapour which the same volume of air could hold at the same temperature.

Runoff: The portion of the precipitation, which is not abosrbed by the soil but finds its way into the streams after meeting the persisting demands of evapo-transpiration, interception and other losses.

Seasonal drought: A kind of drought which occurs during distinct annual periods of dry weather.

Seed bed: A well prepared land for sowing/planting.

Seed hardening: It is a process of subjecting seeds before sowing to alternate cycles of wetting and drying to induce tolerance to drought.

Semi-arid: It describes a land or climate where the annual rainfall is less than 750 mm. and the potential evapo-transpiration is more than precipitation.

Shelterbelt (wind break): A wind barrier of living trees and shrubs established and maintained around crop fields.

Silviculture: It is the art and science of cultivating forest crops as distinguished from silvics which deal with the study of the life history and general characteristics of forest trees of crops.

Silvi-pasture: The practice of growing forage crops in association with forest crop simultaneously on the same land and at the same time. Under the silvi-pastoral system in intervening spaces between the two rows of trees are utilized for growing nutritious grasses/grass legume mixture instead of allowing weeds to grow in its place. It is an intergrated approach of management of land, water and plant resources to fix higher solar energy in less production area.

Silvi-pastoral system: It is a land management system in which forests are managed for the production of wood as well as for the rearing of domesticated animals. In this system, animals are kept and permitted to graze within the forest.

Smother crop: A dense growing crop, which sub presses or stops the growth or a less competitive crop. It is grown for the suppression of weeds.

Soil crusting: A is defined as hardening of soil of more than 5 cm depth. Soil crusting creates soil capping which leads to considerable runoff of rain water. Soil crusting after seeding resulting from soil dispersing rain and sudden drying results in improper and poor emergency of germinating seedlings.

Soil erodibilty factor: The soil erodibilty factors indicates the inherent erodibilty of a soil. It gives an indication of the soil loss form a unit plot 22 mm long with a 9 per cent slope and continuous fallow culture.

Soil moisture characteristic curve: The graphical presentation of the relationship between metric suction and moisture content. In the low suction range 0 to 1 bar, the amount of water retained by the soil depends mainly on the pore size distribution which is a function of the structure of the soil matrix but in the higher suction range the moisture retention is due mainly to the adsorption phenomenon which is a function of soil texture.

Soil moisture deficit: The amount of water needed to return the soil moisture status to the field capacity.

Soil moisture retention curve: The amount of water remaining in a soil at equilibrium is a function of the matrix suction. This function is measured experimentally and represented graphically by a curve know as soil moisture retention curve. The knowledge of moisture retention curves helps in determining the maximum water storage capacity and the available water capacity at different suction values.

Soil moisture stress: It is the sum of soil moisture tension and the osmotic pressure of the soil solution.

Soil moisture tension: It is the force with which the water is held by soil against gravity. It is generally expressed in atmospheres or in mm of mercury.

Strip cropping: (i) Growing soil conserving and soil depleting crops in alternate strips running perpendicular to the slope of the land or to the direction of prevailing winds for the purpose of reducing erosion. (ii) Growing two or more crops simultaneously in different strips wide enough to permit independent cultivation but narrow enough for the crop to interact ergonomically.

Sub humid: A climate or region where the moisture content is below that of under humid conditions but normally adequate for many crops to grow without irrigation.

Stubble mulch: A mulch consisting of the stubble and other crop residues left in and on the surface of the soil as protective cover during the preparation of the seedbed and during the best part of the growing of the succeeding crop.

Suberization: The formation of a complex wax like substance in the cell walls of wounded or cut plant tissue to reduce water loss and protect the area.

Sustainable agriculture: Sustainable agriculture should involve the successful management of resources for agriculture to satisfy changing human needs while maintaining or enhancing the quality of the environment and conserving natural resources.

Temporary wilting: Wilting condition of the plant due to loss of water which would recover from wilted condition during night without addition of any water.

Terrace cultivation: The growing of crops on level steps or terraces that have been constructed on hillsides. Banks or walls are built to hold the soil in place and, where paddy is grown, to enable water to stand on the field. Terrace cultivation takes place where land is scarce.

Terrace: (i) A flat, level or nearly level, narrow, area of land bordering a river to lake, bounded on at least on side by a definite steep slope rising upwards from it, and on the other sides by downward slopes, (ii) A low embankment or ridge of earth constructed across a slope to control surface runoff and minimize soil erosion, (iii) sloping ground cut into a succession of benches and steep inclines are made quite steep and are protected by riprap or retaining walls are substituted, thus giving greater areas for cultivation.

Tillage: The mechanical manipulation of soil for any purpose; but in agriculture it is usually restricted to the modifying soil conditions for crop production.

Tilth: The physical condition of soil as related to its ease of tillage, fitness as seed bed and its impedance to seedling emergence and root penetration.

Turgor potential: Part of total plant water potential arising from the forces exerted on the walls from the water attracted to the cell by the solutes and solids in the protoplast. The turgor potential is also termed as turgor pressure/ wall pressure/ hydrostatic pressure or pressure potential.

Turgor pressure: The pressure set up inside a plant cell due to the hydrostatic pressure of the vacuole contents pressing against the cell wall.

Ultimate wilting point: The moisture content of the soil at which plants growing in that soil are completely wilted and will not recover in an approximately saturated atmosphere without addition of water to the soil. The lower end of the wilting range.

Water balance: The balance between demand and supply of the water for a given place and time, also called inflow outflow and may be represented by the following equation; water balance = ET + surface runoff+ sub surface drainage + change in soil water content.

Water conservation: The physical control /protection /management, and use of water resources in such a way as to maintain crop, grazing and forest land, vegetative cover, wildlife habitat for maximum sustained benefits to people, agriculture, industry, commerce, and other segments of the national economy.

Water deficit: Shortage of water or inadequate water supply. The amount of water needed to return the soil moisture status to the field capacity.

Water harvesting: (i) The practice of collecting water from an area treated to increase run off from rainfall and snow melt precipitation from prepared water sheds for beneficial use, (ii) The collection of runoff rain water from treated or untreated land surface and storing it in open water reservoirs or in the soil itself.

Water holding capacity: The weight of water held by a given quantity of absolutely dry soil when saturated.

Water logging: The saturation of soil with water, causing the water table of soil to rise high enough to expel normal soil gasses and interfere with plant growth or cultivation. It is an major irrigation problem associated with poor drainage. It cause soil degradation.

Water potential: (i) refers to the difference between the activity of water molecules in pure distilled water at atmospheric pressure at 30° C (standard conditions) and activity of these water molecules in any other system. The activity of these water molecules may be greater (positive) or less (negative) than the activity of the water molecules under standard conditions, (ii) the amount by which the chemical potential of water in the system is less than that of the pure water under standard conditions. It enables the prediction of the direction of osmosis and the relative rate of diffusion.

Water regimes: Regime means level; different levels of water (standing or flooding) depth applied in irrigation experiments. Such as 2.5, 5.0, 7.5, 10 cm, etc. standing water to saturation or field capacity to rice crop, etc.

Water saturation deficit (WSD): Parameter for measuring water deficits in plants. It is calculated from the leaves as shown below: WSD = fully turgid wt. - fresh wt. / fully turgid wt. - dry wt) × 100.

Watershed: The total run off from an area which supplies water to a river or lake. A catchment area; a drainage basin.

Water table: The upper boundary for ground water. The upper surface of the locus of points at which the pressure in the ground water is equal to atmospheric pressure.

Waterlogged: (i) Filled or saturated with water or so filled with water as to be heavy or unmanageable, like a completely saturated log, (ii) A condition of land where the ground water stands at a level that is detrimental to plant growth. It may result from excessive irrigation or seepage, coupled with inadequate drainage.

Watershed management: The planned use of watershed lands in accordance with predetermined objectives, such as the control of erosion, stream flow, sedimentation and the improvement of vegetative cover and other related resources.

Waterway: A way or channel for water; a way of water; navigable width of channel.

Weir: (i) A dam in a water course to stop the flow and raise the level, (ii) A dam in a waterway or conduit used to control the water level or the flow.

Wet land: An area of land that has hydric soil and hydrophytic vegetation typically flooded for a part of the year and forming a transition zone between aquatic and terrestrial systems.

Wilt: The collapse of leaves due to unfavorable water relations. It may be due to excessive transpiration, blocking of the xylem elements, pathogens or parasites, a vascular disease that interrupts the normal uptake; or a toxin produced by an organism.

Witling coefficient (WC): The water content of the soil at the time when the leaves of the plant under study first undergo permanent reduction in water supply. Permanent reduction means a degree of wilting from which the leaves cannot recover in an approximately saturated atmosphere unless water is added to the soil. The value is expressed as a percentage of the dry weight of the soil. Its approximate relationship with moisture equivalent, hygroscopic coefficient and moisture holding capacity is given as: WC = moisture equivalent / (1.84 + 0.013), WC = hygroscopic coefficient / (0.68+ 0.012), WC = moisture - holding capacity - 21/2.9 +0.06).

Wilting point: The minimum quantity of water in a given soil necessary to maintain plant growth. When the quantity of moisture falls below this, the leaves begin to drop and shrivel up. In any given soil the quantity is practically constant for all plants, but it increases with a decrease in the size of soil particles. The moisture content of a given soil at which the force holding water to the soil particles equals the maximum water absorbing force of the plant roots.

Wilting range: The range in moisture content of the soil between the permanent wilting point and the ultimate wilting point.

Wind brakes (shelter belts): A strip of trees or shrubs of crop plants serving to reduce the force of wind and provide a protective shelter against wind.

Xerophyte: A plant with structural and physiological features, which permit it to grow in a dry habitat or under water stress conditions.

Zero tillage: A term used when crops are grown only with the tillage of planting (no seed bed preparation), usually possible when all weeds are controlled by more than one application of various herbicides. Also called plough less farming, crop residues planting or slot planting.

◆

FREQUENTLY ASKED QUESTIONS RELATED TO DRYLAND/RAINFED AGRICULTURE

Q. 1. Categorize the dryland agriculture depending on the amount of rainfall received.

Ans. Depending upon the amount of rainfall received dryland agriculture can be grouped into three categories:

i. Dry farming

ii. Dryland farming

iii. Rainfed farming

Q. 2. What is dry farming?

Ans. Dry farming is the cultivation of crops in areas where rainfall is less than 750 mm per annum. Prolonged dry spells during crop period are most common. Crop failures are more frequent under dry farming conditions. Dry farming regions are equivalent to arid regions and moisture conservation practices are important in this region.

Q. 3. What do you mean by dryland farming?

Ans. Cultivation of crops in areas receiving rainfall between 750 and 1150 mm is known as dryland farming. Dry spells during crop period occur, but crop failures are less frequent. Dryland farming areas are grouped under semiarid regions. Adoption of soil moisture conservation practices and provision of drainage especially in black soils are necessary.

Q. 4. What is rainfed farming?

Ans. Cultivation of crops in regions receiving more than 1150 mm rainfall is known as rainfed farming. It is practiced in humid regions where crop failures are rare and drainage is the most important problem.

Q. 5. Give area under irrigated and rainfed farming in India.

Ans. Area under irrigated and rainfed farming in India is as follows:

Class	Area (M ha)	% of total arable land
Total arable	143.8	-
Dryland	34.5	24.0
Rainfed	65.5*	45.5
Irrigated	43.8	30.5

* Includes flood - prone area of 15 m ha.

Q. 6. Give distribution of geographical area in different rainfall zones of India.

Ans. Distribution of geographical area in different rainfall zones of India is as under:

Rainfall zone (mm)	Geographical area (M ha)	Rainwater availability (M ha-m)
100-500	52.07	15.62
500-750	40.26	25.16
750-1000	65.86	57.63
1000-2500	137.24	205.86
>2500	32.57	95.73
Total	328.00	400.00

Q. 7. Give dry farming regions of the world.

Ans. The following are the five arid zones in the world:

(i) North African Eurasian – Sahara and Thar desert

(ii) North American desert – Arizona in USA

(iii) South American desert – Peru

(iv) South African desert – Nambia

(v) Australian desert – Central Australia

The climate of arid region is characterized by very low rainfall, usually less than 200 mm per year, occurring in a very short period. Rainless dry spells, may at times, stretch for more than a year.

Q. 8. What are the effects of rainfall aberrations on crops?

Ans. Effect of rainfall aberration on crops are given as under:

	Rainfall aberration	Effect on crops
(a)	Delay in onset of rainfall	Length of cropping season or cropping duration is reduced, crop sowing is delayed
(b)	Early withdrawal or cessation of rainfall	Moisture stress at maturity, grain filling is affected (terminal stress)
(c)	Intermediate dry spells	
	(i) Immediately after sowing	Germination affected, population reduced.
	(ii) At vegetative phase	Affects stem elongation, leaf area expansion, branching or tillering
	(iii) At flowering	Affects anthesis and pollination grain/pod number reduced
	(iv) At ripening	Grain filling and size reduced

Q. 9. What do you understand by moisture stress ?

Ans. Soil moisture is the most limiting factor in dry land agriculture. Stress is the result of external factors on an organism. Moisture stress indicates the action of lack or excess of water on plants. However, the term moisture stress in dry farming is generally used to imply water deficits.

Q. 10. How moisture stress develops in plant in dry land conditions ?

Ans. Water deficit for moisture stress occurs in the plant, whenever transpiration exceeds water uptake. It may due to excessive water loss through transpiration, less absorption of water or both. Even when there is sufficient soil moisture after irrigation or rain, water deficits develop due to higher transpiration than absorption, especially on hot mid-days. This temporary wilting is known as **incipient wilting** or mid-day depression. As days progress, the soil moisture is depleted and plants can not absorb sufficient amount of water to meet the demand. Water deficits, therefore, occur for prolonged periods. Under such conditions, stomata are partially closed during most of the day except in the morning and evening. Stomatal closure occurs when leaf water potential approaches around -10 to -12 bars in soybean. When the soil moisture reached around -15 bars, several crops show wilting symptoms most of the day but do not die. It is referred as **permanent wilting point.**

Q. 11. How did Koppen classified the climate?

Ans. Koppen has taken annual and monthly means of temperature and also precipitation for classification of climate. Depending on temperature, the climate is divided into tropical, temperate, cold and polar. In tropical climate, even the coolest month has an average temperature (average of minimum and maximum temperature) more than 18°C.Regions with average temperature of coolest month in the range of 18°C to (-)3°C are grouped under temperate climate. Cold climates are those with average temperature of less than (-)3°C in coolest month. If the average temperature of warmest month is less than 10°C, it is called polar climate.

Q. 12. Give the climatic regions of India as classified by Troll.

Ans. Climatic regions of India are as follows:

Climate	No. of humid months	% of geographical area of India
Arid	< 2	17.00
Semi-arid-dry	2.0-4.5	57.17
Semiarid – wet	4.5-7.0	12.31
Humid	> 7.0	1.10

Q. 13. What is moisture deficit index (MDI)?

Ans. Moisture deficit index = $\frac{\text{P-PET}}{\text{PET}}$

Where, P is precipitation and PET is potential evapotranspiration.

Q. 14. Classify the climate on the basis of MDI.

Ans. Based on MDI, the climate is divided into three regions as shown below:

Type of climate	MDI
Sub humid	0 to 33.3
Semiarid	(-) 33.3 to (-) 66.6
Arid	> (-) 66.6

Q. 15. What are the major problems of dry farming?

Ans. The major problems of dry farming are as follows:

(i) Inadequate and uneven distribution of rainfall

(ii) Late onset and early cessation of rains

(iii) Prolonged dry spells during the crop period

(iv) Low moisture retention capacity

(v) Low fertility of soils

Q. 16. Define drought. Also suggest suitable methods for combating the problem of drought.

Ans. Drought constitutes a grave hazard to agriculture in both temperate and tropical conditions. Although there are various definitions of the term drought, there is general agreement that a **drought** can be said to occur whenever the supply of moisture from precipitation or stored in the soil is insufficient to fulfil the optimum water needs of plants. Four types of drought can be identified. These are permanent, seasonal, contingent and invisible droughts.

Drought damage to growing crops can be prevented by:

(i) Grow crops with low water demand.

(ii) Drought resistant short -season crops varieties should be planted.

(iii) Improved agronomical practices which help to conserve soil moisture should be practiced. For instance, growing of legumes and grasses, addition of organic matter or farm residues, timely weed control, adoption of in-situ moisture conservation practices, rain water harvesting and recycling.

Q. 17. Classify drought according to its duration.

Ans. According to duration of drought it can be classified into four categories:

(i) **Permanent drought:** This is characteristic of the desert climate where sparse vegetation growing is adapted to drought and agriculture is possible only by irrigation during entire crop season.

(ii) **Seasonal drought:** This is found in climate with well defined rainy and dry season. Most of the arid and semi-arid zones fall in this category. Duration of the crop varieties and planting dates should be such that growing season should fall within rainy season.

(iii) **Contingent drought:** This involves an abnormal failure of rainfall. It may occur almost anywhere especially in most parts of humid or sub-humid climate.

(iv) **Invisible drought:** This can occur even when there is frequent rains in an area. When rainfall is inadequate to meet the evapotranspiration losses, the result is border line water deficiency in soil resulting in less than optimum yield. This occurs usually in humid regions.

Q. 18. Classify drought on the basis of relevance to the user.

Ans. The classification is given below:

(i) **Meteorological drought:** It is defined as a condition, where the annual precipitation is less than the normal over an area for prolong period (month, season or year).

(ii) **Atmospheric drought:** It is due to low air humidity, frequently accompanied by hot dry winds. It may occur even under condition of adequate available soil moisture.

(iii) **Hydrological drought:** Meteorological drought when prolonged results in hydrological drought with deplection of surface water and consequent drying of reservoirs, tanks, etc.

(iv) **Agricultural drought:** It is the result of soil moisture stress due to imbalance between available soil moisture and evapotrans-piration of a crop.

Q. 19. What do you understand by escaping drought?

Ans. Evading the period of drought is the simplest means of adaptation of plant to dry conditions. Many desert plants so called ephemerals germinate at beginning of rainy season and have an extremely short life period (5 to 6 weeks) which is confined to rainy period. These plants complete their life cycle before moisture stress is noticed. They do not have any mechanism for overcoming moisture stress and are, therefore, not drought resistant.

Q. 20. What are the crops and varieties which may act as ephemeral?

Ans. In cultivated crops, the ability of a cultivar to mature before the soil dries is the main adaptation to grow in dry regions. However, only very few crops have such a short growing season to be called as ephemeral.

Certain varieties of pearlmillet mature within 60 days after sowing. Short duration pulses like cowpea, greengram, blackgram can be included in this category. In addition to earliness, they need drought resistance because there may be dry spells within the crop period of 60 days.

Q. 21. What do you mean by drought resistance?

Ans. Plant can adopt to drought either by avoiding stress or by tolerating stress due to different mechanisms. These mechanisms provide drought resistance.

Q. 22. What do you mean by avoiding stress?

Ans. Stress avoidance is the ability to maintain a favourable water balance and turgidity even when exposed to drought conditions, thereby avoiding stress and its consequences. A favourable water balance can be achieved by (i) conserving water by restricting transpiration behavior or as soon as stress is experienced; or (ii) accelerating water uptake sufficiently so as to replenish the lost water.

Q. 23. What mechanisms exist in plants for water conservation?

Ans. The following mechanisms exists in plants for water conservation:

(i) **Stomatal mechanism:** Drought resistant varieties are capable of regulating the stomatal mechanism and closing stomata when drought prevails. In these plants photosynthesis can proceed with the least loss of water.

(ii) **Increased photosynthetic efficiency:** When stomata are closed as a mechanism against water loss, photosynthesis is affected due to reduction in entry of carbon dioxide through stomata. Some plants overcome this problem by their increased photosynthetic efficiency. In C4 plants, the carboxylating enzyme namely phosphoenolpyruvic acid carboxylase has very high affinity for carbon dioxide and high potential activity. Therefore, for the same amount of stomatal opening, C4 plants have higher photosynthetic rate than C3 plants. Thus, C4 plants are said to be drought resistant as they are able to grow better even under moisture stress. In addition, C4 plants translocate photosynthates more rapidly. The C4 plants like sorghum, maize, pearlmillet, sugarcane, *etc.* can photosynthesize even when carbon dioxide concentration is 10 ppm while in most plants photosynthesis stops when carbon dioxide concentration is about 50 to 100 ppm.

The plants with crassulacean acid metabolism (CAM) are highly drought resistant. They open stomata only during night when they take carbon dioxide into the leaves and store it as organic acid. This stored organic acid is released inside the leaf and utilized in photosynthesis. By this mechanism, the amount of water lost is minimum without reduction in photosynthesis. The CAM type plants are pineapple and agave.

(iii) **Lipid deposits on leaves:** Some plants like soybean, sorghum, etc. reduce water loss by depositing lipids on plant surfaces under moisture stress.

(iv) **Reduction in leaf area:** Most of the plants reduce transpiration by reducing leaf area in moisture deficit conditions.

(v) **Leaf surface:** Leaves with thick cuticle waxy surface and spines, etc. are effective in reducing transpiration.

(vi) **Effect of awns:** Awned varieties give more yield under drought conditions compared to awnless varieties. Awns contribute 12 per cent of photosynthesis to grain.

(vii) **Water storage in plants:** Water stored in plants is considered an adaptive mechanism. However, its significance is small in crop plants. The leaves of pineapple contain substantial amounts of water in special non-chlorophyllous tissues which is utilized during drought.

Q. 24. What mechanisms exist in plants that improve water uptake?

Ans. Water uptake can be improved by several mechanisms as:

(i) **Efficient root system**- Deep, well branched and rapidly growing root system helps in absorbing more moisture by exploiting higher volume of soil. This helps in avoiding drought. Variation in root length between varieties is related to differences in drought resistance.

(ii) **Root-shoot ratio**- If the roots are more compared to transpiring shoots, water balance can be maintained.

(iii) **Increase in liquid phase conductance**- To maintain high water potential in plants, not only uptake is important, but also conductance. Lowering of resistance to water can be achieved by increasing either diameter of xylem vessels or their number.

(iv) **Osmotic adjustment**- Stress induced break down of carbohydrates and proteins increase the concentration of solutes in cell sap resulting in reduction of osmotic potential.

Q. 25. What is difference between drought avoidance and drought tolerance?

Ans. Due to different drought avoidance mechanisms, plants are able to maintain favourable water balance and adverse effects of reduced water potential are not felt by the plants. While in drought tolerance, water potential of plant is reduced and its adverse effects are felt.

Q. 26. What do you mean by drought tolerance?

Ans. **Drought tolerance** can be defined as tolerance of plants to a level of stress at which 50 per cent of cells die. The performance of higher plants depend upon the integrated function of many cells which is disturbed by drought.

Q. 27. What mechanisms of drought tolerance exists in the plants?

Ans. Drought tolerance in plants is due to two important mechanisms:

(i) **Mitigating stress** - The simplest way of mitigating stress is by resisting dehydration and by maintenance of higher osmotic pressure by accumulating higher amounts of solutes. The leaves with thick cuticle resists cell collapse.

(ii) **High degree of tolerance**- The death of the plants can be avoided either by reducing metabolic strain or plastic strain during drought. Death of the cells occur either due to reduction in the carbon metabolism (photosynthesis) or nitrogen metabolism (protein synthesis). When plants are subjected to severe moisture stress, stomata are closed permanently resulting in cecession of photosynthesis and starvation of plants. However, plants capable of keeping the stomata partially open can photosynthesize and survive. As a consequence of drought, proteins are broken down into amino acids. Protein break down may result in accumulation of products to a toxic level. In drought resistant plants, the net loss of RNA is prevented and the plants are capable of continuous protein synthesis. Young leaves are more resistant to drought than older leaves due to higher protein content. Plastic strain indicates irrevocable loss of plant tissues due to severe stress. The mechanism involved is killing of transpiring portions and keeping propagation material without desiccation. In many species of perennial plants, the above ground parts die off during drought and underground parts such as rhizomes, bulbs, tubers, etc. remain alive but dormant.

Q. 28. Describe the desirable plant characteristics for drought resistant crop in brief.

Ans. The desirable characteristics are as follows.

(i) Rapid germination and early establishment of deep roots.

(ii) Rapid phenological development.

(iii) Parahelonostic movements.

(iv) Development plasticity.

(v) Stomata sensitive only to large vapour pressure deficits and insensitive to low leaf water potential.

(vi) Ability to adjust osmotically.

(vii) Large transfer of assimilates from stem to grain.

(viii) Dehydration tolerance particularly at seedling and grain filling stages.

(ix) Efficient root system, capacity to grow in dry soils.

(x) High rate of photosynthesis with thick small leaves.

Q. 29. What do you mean by water harvesting?

Ans. Collecting and storing water on the surface of the soil for subsequent use is known as water harvesting. It is method to induce, collect, store and conserve local surface runoff for agriculture in arid and semi-arid regions.

Q. 30. What are the major differences in water harvesting in arid and semi arid regions?

Ans. Water harvesting is done both in arid and semi-arid regions with certain differences. In arid regions, the collecting area or catchment area is substantially in higher proportion compared to command area. Actually, the runoff is induced in catchment area in arid lands whereas in semi-arid regions, runoff is not induced in catchment area only the excess rainfall is collected and stored.

Q. 31. What are the important methods of water harvesting in arid regions?

Ans. The different methods of water harvesting followed in arid regions are as follows:

(i) **Runoff Farming:** In this method the watersheds are divided into small catchment areas of one to three hectares that allowed runoff water to be collected in easily constructed channels on the hill sides and are small enough to prevent uncontrollable amounts of water. The channels that lead the water to cultivated fields are terraced and have stone spillways so that surplus water in one field could be led to lower ones. Constructed small check dams with rocks across the small gullies and guided the water to fields.

(ii) **Water Spreading:** In arid areas, the limited rainfall is received as short intense storms. Water swiftly drains into gullies and then flows towards the sea. Water is lost to the region and floods caused by this sudden runoff can be devastating often to areas otherwise untouched by the storm. Water spreading is a simple irrigation method for use in such a situation. Flood waters are deliberately diverted from their natural courses and spread over adjacent plains. The water is diverted or retarded by ditches, dikes, small dams or brush fences. The wet flood plains or valley floods are thereby used to grow crops.

(iii) **Micro Catchments:** A plant can grow in region with too little rainfall for its survival if a rain water catchment basin is built around it. The basin stores the runoff from micro catchment.

Q. 32. What water harvesting techniques are followed in semi-arid regions?

Ans. Water harvesting techniques followed are as follows:

(i) **Dug Wells:** Hand dug wells have been used to collect and store underground water and this water is lifted for irrigation. The quality of water is generally poor due to dissolved salts.

(ii) **Tanks:** Runoff water from hill sides and forests is collected on the plains in tanks. The traditional tank system has components viz. catchment area, command area, storage tank, tank bund, sluice and spillway. The runoff water from catchment area is collected and stored in storage tank on the plains with the help of a bund. To avoid the breaching of tank bund, spillways are provided at one or both the ends of the tank bund to dispose of excess water. The sluice is provided in the central area of the tank bund to allow controlled flow of water into the command area. The command area of many tanks ranges from 25 to 100 hectares. In areas receiving annual rainfall of about 1000 to 1500 mm with runoff around 50% where canal and well irrigation is not feasible due to topography and underground water table, tanks are found suitable. An earthen dam of 12 m height is constructed to store about 55,600 m^3 of water from a forest watershed of 10 hectares. This tank can provide supplimental irrigation for 20 hectares of rainfed farm land. However, there are only few tanks with command area of more than 1,000 hectares. Unlike wells, the quality of water is good in tanks. Water from the tanks is used to irrigate the command area by gravity flow.

(iii) **Farm ponds:** Farm ponds of size 100 to 300 m^3 may be dug to store 30 per cent of runoff. The problem associated with farm ponds is high seepage loss. This can be reduced by lining of walls. Some of the traditional methods for seepage control are the use of bentonite, soil dispersant and soil cement mixture. Bentonite has excellent sealing properties if kept continuously wet, but cracks develop when dried. Soil-cement mixture can be used, but surface cracking develops when exposed to sun drying. A soil - cement lining of 100 mm thickness reduces seepage losses up to 100 per cent. The pit lined continuously develops cracks but no cracks develop when applied in blocks. The other alternative sealant for alfisols is mixture of red soil and black soil in the ratio of 1:2.

(iv) **Percolation tanks:** Flowing rivulets or big gullies are obstructed and water is ponded. Water from the ponds percolates into the soil and raises the water table of the region. The improved water level in the wells surrounding the percolation tanks are used for supplemental irrigation.

(v) **Inter-row water harvesting:** In areas of high rainfall, there is possibility for occasional waterlogging and yields of crops like maize is affected. Growing maize on beds and rice in furrow helps in increasing the yield of both the crops. The excess water collected on beds is stored in furrows which is beneficial for rice.

(vi) **Broad bed and furrows (BBF) and Farm ponds:** The seed bed is made into wide beds of 1.20 m alternated with the furrows of width 0.30 m. These are called Broad Bed & Furrows). This method is quite suitable for black soils when combined with dry sowing. Excess water during rainy season is collected through the furrows and passed through grassed waterways into farm pond where the water can be stored for subsequent use.

Q. 33. What measures can be taken to check loss of water through evapo-transpiration?

Ans. In dryland agriculture, soil moisture is the most limiting factor, it is lost as evaporation from the soil surface and as transpiration from the plant surfaces. Evaporation has to be arrested as it is not directly related to productivity whereas transpiration can be reduced to some extent without affecting productivity of plants. The evapo-transpiration losses can be reduced by:

(i) Mulches (ii) antitranspirants

(iii) wind breaks and (iv) weed control

Q. 34. What are the advantages of cover cropping ?

Ans.

Advantages:

- It reduces runoff and conserve moisture
- It prevents soil erosion, increases water infiltration and water holding capacity of soil.
- Cover crops add organic matter to the soil and improve soil tilth.
- It protects newly constructed terraces and other mechanical structures used for the control of soil erosion.
- Cover crops provide green fodder to the cattle and improve soil fertility.

Q. 35. Give cultural methods for transpiration control.

Ans. Cultural methods for transpiration control are given as under :

(a) **Weed control:** Most of weeds have a high transpiration coefficient i.e. amount of water transpired to produce unit quantity of dry matter. Early weed control prevents unwanted transpiration loss of water through weeds.

(b) **Shelterbelts:** Rows of trees grown across the direction of wind reduce air movement, reduce temperature of air and plant canopy, increase humidity in the protected strips and thereby reduce the atmospheric water demand and control transpiration in the inter space between shelterbelts.

(c) **Alley cropping:** In this arable food crops are grown in the alleys formed by hedgerows of perennial trees or shrubs. A similar effect on reduction in atmospheric water demand and transpiration as described under shelter belts is caused in alley cropping. This method is also called as hedge row intercropping. The best plants for this are subabul and *Glyricidia* because they grow well in degraded soils. In the hedge rows of subabul, blackgram, cowpea, sunflower or groundnut can be intercropped.

Q. 36. What do you mean by antitranspirant? Also give the types of antitranspirants used in agriculture?

Ans. Nearly 99 per cent of water absorbed by the plants is lost in transpiration. If transpiration is controlled, it may help in maintenance of favourable water balance in the plant system. **Antritranspirant** is any material applied to transpiring plant surfaces for reducing water loss from the plant.

There are four types of antitranspirants.

(i) Stomatal closing type — Phenyl Mercuric Acetate (PMA), atrazine, etc.

(ii) Film forming type — Mobileaf, Hexadicanol, Silicon

(iii) Reflectat type — Kaolin

(iv) Growth retardants — Cycocel

Q. 37. How wind breaks and shelterbelts are useful in checking evaporation losses?

Ans. Wind breaks are any structures that obstruct flow and reduce wind speed while shelterbelts are rows of trees planted for protection of crops against wind. The direction from which wind is blowing is called **windward side** and direction to which wind is blowing is called **leeward side.** Shelterbelts are planted across the direction of wind. They do not obstruct the wind flow completely. Depending upon their porosity, certain amount of wind passes through the shelterbelts while the rest deflects and crosses over the shelterbelts. It thus reduces wind speed without causing turbulence. The protection offered by the shelterbelts is dependent on the height of central tree row in shelterbelts. Generally, shelterbelts give protection form desiccating winds to the extent of 5 to 10 times of their height on windward side and up to 30 times on leeward side. Due to reduction in wind speed, evaporation losses are reduced and more water is available for plants. The beneficial effect of shelterbelts are seen more clearly in drought years. In addition, shelterbelts reduce wind erosion.

Q. 38. How water use efficiency in dryland conditions can be improved?

Ans. Water use efficiency is mass / dry matter produced per unit of water used. It is expressed in various ways such as kg or tonnes of dry weight produced per cm of water evapotranspired or as a simple ratio of kilogram of dry weight per kilogram of evapotranspiration. Following agronomic practices can be used to improve WUE

(i) Practices for reducing evaporation from soil surface viz., mulching, modifying plant population and spacing, selecting varieties with rapid early growth, early sowing and application of fertilizers.

(ii) Practices for increasing water supply to plants viz., rain water harvesting, supplementary irrigation, cultivation to improve infiltration, reduce runoff, weed control, application of fertilizers, multiple and relay cropping and selecting varieties with deeper roots.

Q. 39. What may the possible aberrant weather situations in dry land agriculture?

Ans. Aberrant weather in dry land agriculture are: (i) delayed onset of monsoon (ii) dry spell immediately after sowing (iii) short or long dry spell during growing season in mid season or late season (iv) early withdrawal of monsoon.

Q. 40. What contingent plan would you suggest to farmers under delayed on set of monsoon?

Ans: (i) Sowing under dry condition (ii) raising nursery and transplanting (iii) changing plant population and planting geometry (iv) selection of right crops and varieties depending on the onset of monsoon.

Q. 41. How dry spell immediately after sowing is managed under dryland agriculture?

Ans. The suggested measures are as follows:

(i) **Resowing:** When there is a severe dry spell and the population is too poor to expect any reasonable yield, resowing of the entire field using same or new crop using subsequent rains is necessary.

(ii) **Gap filling:** If the dry spell followed by sowing results in sub-optimum plant stand, with the subsequent showers gap filling may be considered advantageous. If the sown crop is suitable for transplanting and the stage of seedling is adequate for transplanting, transplanting in the gaps using excess seeding growing at some places in the field with subsequent rains e.g. fingermillet and pearmillet should be preferred. If the break in the rain is rather long and do not permit transplantation of already sown crop with existing seedings due to age, other crops grown in region can also be used for transplanting.Similarly reseeding can also be considered in the gaps using crops suited for late planting.

Q. 42. Write contingent crop planning if there is break in monsoon.

Ans. Following practices have been developed to mitigate the damaging effect of dry spells.

(i) **Thinning of the stands:** Thinning of alternate plants in the rows in the event of mild stress and every third row under higher moisture condition is recommended.

(ii) **Removal of weak seedlings:** In most of the dryland crops there may not be uniform emergence because of soil variation, differential depth of sowing or inadequacy of soil moisture at sowing. In most of the cereals the late germinated seedlings will not be vigorously growing and may not be competitive to give any yields. At severe drought conditions these weak seedlings may be removed.

(iii) **Repeated and deep cultivation:** To prevent soil cracking and to break the capillary loss, early and frequent inter cultivation are needed. Generally blade hoes are employed which penetrate hardly to a depth of 5 to 7.5 cm.

(iv) **Mulching:** Mulches have a definite controlling effect on evaporation of soil moisture and hence play an important role in dryland agricultural under field conditions, surface soil is just stirred and soil becomes air dry.

(v) **Stripping of older leaves:** Older leaves are shaded, they transpire more than helping in photosynthesis. So to avoid moisture loss these older leaves may be stripped, which helps in active growth of new parts.

(vi) **Returning ratooning:** In case of pearlmillet and sorghum when crop is exposed to dry spell of 30-35 days during peak month, yields are drastically reduced. Under such conditions cut/harvest the crop for fodder purpose and allow the ratoon with the onset of rains. For quick growth supplemental addition of fertilizer is suggested.

(vii) **Life saving irrigation:** If there are farm ponds/seasonal streams, life saving irrigation should be given to save the crop.

(viii) **Use of antitranspirants:** The possibilities of reducing transpiration is of greatest importance in dry farming. Any material that is applied to transpiring plant surfaces for reducing water loss is called **antitranspirant.** Many herbicides, fungicides and growth hormones posses antitranspirant characteristics.

(ix) **Harvest the crop for fodder:** If there is complete failure of rains and crop is not in position to produce seed yield especially of the *kharif* crop, such crop may be harvested for fodder and the land could be kept ready to make use of shower for post rainy season crops like chickpea, mustard, toria, etc.

Q. 43. Give advantages of seed hardening.

Ans. Advantages of seed hardening are as follows:

(i) Low cost technology, improved seed viability and ensures early germination.

(ii) Helps in better development of root system which helps in absorption of more water and nutrients from deeper layer.

(iii) Germination and seedling emergence are completed before surface soil dries out.

(iv) It induces drought tolerance by increasing the resistance to protoplasmic dehydration in young seedlings subjected to moisture stress.

(v) Relative water content in leaves is improved during the moisture stress conditions.

(vi) Most important requirement for pre-monsoon sowing.

Q. 44. What are requirements of pre-monsoon dry seeding?

Ans. The requirements are as follows:

- It is recommended for bold seeded crops like cotton and sorghum only.
- Time of advance sowing must be fixed based on rainfall analysis for date of onset of monsoon and continuity of rainfall after sowing.
- Seeds must be hardened to ensure quick germination and drought tolerance.
- Seeding depth must be such that seeds will germinate only after receipt of rainfall to wet that depth.
- Surface sowing may lead to germination with less rainfall and death of seedlings due to subsequent soil drying.
- Off season tillage is necessary to enable sowing in dry soil before monsoon.
- Seed damage by soil insects has to be prevented.

Q. 45. What is a watershed?

Ans. **Watershed** can be defined as a topographic, hydrological unit draining runoff water at a common or particular point by a network of channels and streams. Thus, a watershed represents a hydrological unit of area but can also be described as a bio-physical and socio-economic unit for planning and management of natural resources.

Q. 46. What is watershed management?

Ans. **Watershed** is an area of land and water bounded by a drainage divide within which the surface runoff collects and flows out of the area through a single outlet into river or other body of water. In simple words, it can be referred to the divide separating one drainage basin from another. It is also used as a synonym for **catchment area or a drainage basin.**

The intensification of land use in the traditional agricultural setting is self defeating because it is exploitive. The present agricultural practices greatly increase runoff and soil erosion, reduce ground water recharge, cause floods and sedimentation of reservoirs, etc. As a result, the cultivated land resource base is shrinking and its productive capacity is diminishing.

Runoff, erosion and drainage represent serious problems in many areas of semi-arid tropics. These problems can be solved by evolving developmental pattern of the land.The collection of excess water and its utilization to provide greater stability to rainfed agriculture appears to be viable developmental alternative. The watershed is the natural framework for resource development in relation to crop production.

Q. 47. Give the needs and advantages of watersheds.

Ans. Need and advantages of watershed are as follows:

(i) Watershed is an acceptable basic hydrological unit of planning for optimum use and conservation of soil and water resources.

(ii) Here, every part of land including barren, sloppy and marginal lands being treated according to its capability. By adopting watershed as unit for development, different measures are adopted and executed in each of the topo-sequence according to its capability, providing an integrated treatment of arable and non-arable lands.

(iii) It aims at comprehensive development of all resources in the watershed i.e. holistic. It starts from the most important resources viz., soil and water and extends to other resources like crops trees, livestock, etc.

(iv) Some of the resource conservation measures may have to be carried out cutting across field boundaries e.g. contour bunding, contour vegetative barriers, shelterbelts, drainage channel. For this watershed is more ideal unit.

(v) It is a multidisciplinary approach involving scientists from all related disciplines of agronomy, engineering, horticulture, forestry extension, and economics, etc.

(vi) It provides for involvement of farmers in planning, execution and monitoring of the development.

Q. 48. What is the basic purpose of integrated watershed management in dry lands?

Ans. The purpose of integrated watershed management is to practice improved and scientific land management programme for improving the hydrologic behaviour of the watershed by reducing runoff and soil loss. In other words, the watershed development programme aims at preventing degradation of land and increasing land productivity and simultaneously it maintains ecological balance between land, water, plant and animal kingdom in the area.Thus, watershed management particularly in dry lands is regarded as holistic approach towards optimising the integrated use of land, water and vegetation in an area for providing an answer to alleviate drought, moderate floods, prevent soil erosion, improve water availability and increase fuel, fodder, fibre and agricultural production on a sustained basis.

Q. 49. How watersheds are categorized?

Ans. Small watersheds are those where the overland flow is chief contributor to the peak flows and where the overland flows are not affected by the channel characteristics. On large watershed, on the other hand, the channel flows are highly pronounced and peak flows are influenced by the channel characteristics. The area to be managed for soil and water conservation programme is also important in categorizing a watershed.

Thus, these are sub divided in to different levels called as macro, sub-milli, micro and mini watersheds managing area of >50,000, 10,000-50,000, 1000-10,000, 100-1000 and 1-100 ha, respectively.

Q. 50. Mention aspects of watershed management that are especially important on steep hill slopes.

Ans. For steep hill slopes the following aspects are to be considered:

(i) Ensuing adequate protection of land against soil erosion with the use of local resources and skill based soil conservation measures.

(ii) Maximum retention of rainfall within the area without affecting the crops.

(iii) Shortage of runoff water, its use for pisciculture and irrigation.

Q. 51. Why is vetiver considered best for building a live terrace?

Ans. The deep-rooted vetiver plants possess dense stiff stems that tend to filler out soil particles from the muddy runoff. This sediment builds up on the upslope side of the grass barrier and in time actually creates a terrace that is often more than 1 m above the soil surface on the down slope side of the plants.

Q. 52. What steps to be taken in integrated watershed management to get maximum production and stabilize rural incomes under dryland agriculture?

Ans. In dry farming areas, watershed approach provides an ideal means for integrated development. This approach will further the cause of sustainability which requires the best in land water use and ecological sustenance. There is need for a multipronged approach to maximize crop production and also ensuring stability in rural incomes. The following are the various steps:

(i) Crop production assumes importance in watersheds. It is essential to practice *in-situ* moisture conservation and water harvesting.

(ii) In crop production, priority is for intercropping.

(iii) Crop substitution, depending on suitability of soil, can ensure higher returns.

(iv) Site specific crop production techniques have to be adopted.

(v) Afforestation component has to be planned carefully and implemented vigorously so as to provide for the farmer affordable access to fuel for cooking, timber for building and wood for implements. Cultivable wastes and marginal lands should be used for this purpose.

(vi) Dryland horticulture will help to minimise risk and offer scope to improve returns particularly from marginal lands. Mango, guava, ber, pomegranate, sapota and tamarind have good potential.

(vii) Discouraging raising of rice crop under existing irrigation facilities as rice requires large quantities of water especially in summer and using the water saved for extending the area under irrigation.

(ix) More efficient exploitation and use of ground water, coupled with conjunctive use of rain water and using it for high value horticultural crops like grape, orange, vegetables and flowers.

(x) Encouraging dairying and adequate support systems such as fodder, range, pasture, etc.

(xi) Stall-fed goat and sheep farming needs to be popularized.

(xii) Sericulture and mulberry cultivation and bee keeping has to be intensified.

(xiii) Processing of farm and horticultural produce into value added products offers scope for off-season occupation besides ensuring higher returns to the farmer.

(xiv) Custom hire services at the village level such as ploughing, plant protection, harvesting, etc. will provide opportunities for extended employment.

(xv) The extension and training programmes should equip the farmers to respond to the emerging complexities of agriculture. The training of young farmers has to focus on the consequences and limits as new technology in relation to ecological and environmental concerns.

Q. 53. What are the principles of watershed management?

Ans. The main principles of watershed management based on resource conservation, resources generation and resource utilization are:

(i) Utilization of land as per its capability class.

(ii) Protection of productive soil.

(iii) Reduction of siltation hazards.

(iv) Maintain adequate vegetation cover on the surface of soil throughout the year.

(v) *In-situ* conservation of rain water.

(vi) Safe diversion of excess water to storage points.

(vii) Stabilization of gullies and increasing ground water recharge.

(viii) Increase cropping intensity and land equivalent ratio (LER) through intercropping and sequence cropping.

(ix) Safe utilisation of marginal lands through alternate land use system with agriculture-horticulture-forestry-pasture systems with different options and combinations.

(x) Water harvesting for supplemental and off season irrigation.

(xi) Maximization of productivity per unit area per unit time and per unit of water.

(xii) Ensuring sustainability of the eco-system befitting the man-animal-plant-water complex.

(xiii) Maximization of combined income from inter related and dynamic crop-livestock–tree-labour income and to cut down risks during aberrant weather situations.

(xiv) Improving the socio-economic status of the farmers.

Q. 54. Describe 'Power' as symbolic of objectives of watershed management.

Ans. **"Power".** The letters symbolizes the following:

P = Production of food–fodder-fuel–fruit-fibre-fish-milk combine on sustained basis.

- pollution control
- prevention of floods

O = Over exploitation of resources to be minimized by controlling excessive biotic interferences like over grazing.

- Operational practicability of all on-farm operations and follow up programmes including easy approachability to different locations in watershed.

W = Water storage at convenient locations for different purposes.

- Wild animal and indigenous plant life conservation at selected places.

E = Erosion control

- Eco-system safety
- Economic stability
- Employment generation

R = Recharge of ground water

- Reduction of drought hazards
- Reduction of siltation in multi purpose resources
- Recreation.

Q. 55. What are the components of watershed management programme?

Ans. The main components of watershed programme are

(i) Soil-water conservation and water harvesting

(ii) Crop management

(iii) Alternate land use system

Successful implementation of the above these components in the watersheds would lead to the development of livestock, poultry, pisciculture and other associated activities ultimately resulting in improved standards of living of the farm families.

Q. 56. Enlist the advantages of broad bed furrow (BBF) system.

Ans. Advantages of BBF system are as follows:

(i) It can accommodate a wide range of crop geometry i.e. close as well as wide row spacing.

(ii) It is suitable for both sole cropping and intercropping systems.

(iii) Furrows serve to safely guide runoff water in the early part of rainy season and store rainwater in the later stages.

(iv) Sowing can be done with seed drills.

(v) It can be formed by bullock drawn or tractor drawn implements. Bed former cum seed drill enables BBF formation and sowing simultaneously, thus is reducing the delay between rainfall receipt and sowing.

Q. 57. Give the advantages of farm ponds.

Ans. Advantages of farm ponds are as follows:

- Harvested water can be used for protective or supplemental irrigation to crops at critical stages.
- Since runoff is properly guided through grassed waterways, erosion is checked.
- Earth excavated from ponds can be used for bunding and levelling of fields.
- Stored water can be used as drinking water for humans and animals, for spraying of plant protection chemicals and for fish rearing.
- High value tree crops can be raised near farm ponds with protective irrigation.
- A chain of farm ponds can recharge ground water in the region.

♦

OBJECTIVE QUESTIONS RELATED TO DRYLAND /RAINFED AGRICULTURE

I. Multiple Choice Questions

1. The word drainage basin is synonyms of

(a) Watershed (b) Catchment area
(c) Both (a) & (b) (d) None of these

2. The word 'watershed' was introduced in

(a) 1920 (b) 1930 (c) 1940 (d) 1950

3. Land which contributes runoff to a common out let is called

(a) Farm pond (b) Watershed
(c) Nadi (d) None of these

4. National Watershed Development Programme for Rainfed Agriculture was started in

(a) 1970-71 (b) 1977-78 (c) 1986-87 (d) 1990-91

5. The optimum thickness of Low Density Poly Ethylene (LDPE) film for lining of water pond is

(a) 5-10 micron (b) 20-50 micron
(c) 200-250 micron (d) 800-1000 micron

6. Area under dry land agriculture in India is approximately

(a) 75% (b) 65% (c) 55% (d) 45%

7. Contribution of dry land agriculture to total food grain production is around

(a) 44% (b) 50% (c) 70% (d) 80%

8. Which is the stomatal closure type of antitranspirant?

(a) PMA (b) Mobileaf (c) Silicon (d) All of these

9. All India Coordinated Research Project for Dry Land Agriculture (AICRPDA) was started in

(a) 1970 (b) 1980 (c) 1990 (d) 2000

10. Central Arid Zone Research Institute is located at

(a) Ludhiana (b) Kanpur (c) Jodhpur (d) Solapur

11. International Centre for Agricultural Research in the Dry Areas (ICARDA) is located at

(a) Syria (b) Philippines(c) Bangalore (d) Hyderabad

12. A mini watershed has size

(a) 1-10 ha (b) 1-100 ha

(c) 100-1000 ha (d) 1000-10,000 ha

13. The cultivation done in adverse weather conditions is known as

(a) Relay cropping (b) Mixed farming

(c) Contingent cropping (d) Mixed cropping

14. In dry land areas there would be

(a) High root to top ratio (b) High root to leaves ratio

(c) Low root to top ratio (d) Low root to leaves ratio

15. Moisture deficit Index is calculated by formula

(a) $\frac{P - PET}{PET}$ (d) $\frac{PET - P}{PET}$

(c) P - PET (d) PET - P

16. Central Research Institute for Dry land Agricultural (CRIDA) is located at

(a) Hyderabad (b) Jodhpur

(c) Allahabad (d) Bangalore

17. CRIDA was established in

(a) 1980 (b) 1985 (c) 1990 (d) 1995

18. ICRISAT was established in

(a) 1970 (b) 1971 (c) 1972 (d) 1973

19. Cultivation of crops in regions receiving more than 1150 mm rainfall is known as

(a) Rainfed farming (b) Dry land farming

(c) Natural farming (d) Mixed farming

20. Of the following plants which are highly drought resistant ?

(a) CAM plant (b) C3 plants

(c) C4 plants (d) None of these

21. Which is film forming type of antitranspirant ?

(a) Mobileaf (b) PMA

(c) Kaolin (c) None of these

22. The India Meteorological Department defined drought as an area receives less than.........% of the total normal South West monsoon

(a) 50 (b) 75 (c) 60 (d) 80

23. Commercially used chemical for seed hardening for drought tolerance in chickpea is

(a) Potassium dihydrogen phosphate

(b) Potassium chloride

(c) Succinic acid

(d) None of these

24. The size of the water harvesting pond for one ha area is kept

(a) 100-200 m^3 (b) 250-300 m^3
(c) 500-1000 m^3 (d) 800-900 m^3

25. Which of the followings is accumulated in plant body during water stress?

(a) Fats (b) Carbohydrates
(c) Proline (d) None of these

26. Which of the followings will be affected first under water stress conditions?

(a) Cell division and elongation (b) Photosynthesis
(c) Transpiration (d) Protein synthesis

27. The International Centre for Agriculture Research in Dry Areas (ICARDA) was established in

(a) 1987 (b) 1977 (c) 1967 (d) None of these

28. The dryland areas can be designated where

(a) Annual precipitation is equal to potential evapotranspiration
(b) Potential evapotranspiration is more than annual precipitation
(c) Potential evapotranspiration is less than annual precipitation
(d) None to these

29. "Water harvesting" refers to

(a) Collection of runoff water into micro or macro reservoirs and recycling the stored water for irrigation
(b) Conservation of rain water in soil by preventing the runoff so that rain water is better utilised in dryland
(c) A technique of better harvesting the ground water to suppliment the canal irrigation
(d) A method of preventing percolation loss of water by suitably applying moisture barrier substances like asphalt

30. For water harvesting by contour bunds, the cross section of bund is generally kept as

(a) 0.5 m^2 (b) 1.0 m^2 (c) 1.5 m^2 (d) 2.0 m^2

31. Graded bunding is done in

(a) Heavy textured soils (b) Light textured soils
(c) Medium textured soils (d) None of these

32. The construction of dugout farm pond is found suitable for the area where land slope is

(a) $< 3\%$ (b) 10% (c) 15% (d) 20%

33. The spacing of contour lines for cultivation varies from

(a) 10 to 15 m (b) 24 to 40 m
(c) 50 to 75 m (d) 100 to 150 m

34. Engineering measures for soil and water conservation in arable lands are generally adopted for the slope

(a) > 0.2 % (b) > 1.0 %
(c) > 2.0 % (d) None of these

35. The width of buffer strip crops usually varies from

(a) 2 to 4 m (b) 5 to 10 m
(c) 10 to 15 m (d) 15 to 20 m

36. Width of wind strip crops is kept maximum in

(a) Clay loam soil (b) Sandy clay
(c) Sandy loam soil (d) Clay soil

37. Contour bunding is done in

(a) Heavy textured soils (b) Light textured soils
(c) Medium textured soils (d) None of these

38. In splash erosion the raindrop energy acts in the form of

(a) Kinetic energy (b) Potential energy
(c) Chemical energy (d) None of these

49. Accelerated erosion is caused by

(a) Water (b) Wind
(c) Both a and b (d) None of these

40. Sand dune is associated with

(a) Wind erosion (b) Chemical erosion
(c) Water erosion (d) None of these

41. Shelterbelts are constructed by

(a) One row (b) Two rows
(c) More than two rows (d) None of these

42. Stubble mulching is associated with

(a) Water erosion (b) Wind erosion
(c) Glacial erosion (d) None of these

43. Contour bunds are recommended for

(a) Low rainfall areas (b) Medium rainfall areas
(c) High rainfall areas (d) None of these

44. Graded terraces are used in the areas of

(a) Medium to high rainfall (b) High rainfall only
(c) Medium rainfall only (d) None of these

45. The area lost in bench terracing depends on

(a) Slope (b) Depth of cut
(c) Depth of fill (d) Both (b) & (c)

46. Contour trenching is done mainly in
(a) Low rainfall areas for moisture conservation
(b) High rainfall areas for moisture conservation
(c) Arid zones for wind erosion control
(d) All of these

47. Vertical interval of bench terrace depends on
(a) Land slope (b) Soil depth
(c) Both a and b (d) None of these

48. Bench terraces are constructed at slope gradient of
(a) > 2 % (b) > 5 % (c) > 10 % (d) > 15 %

49. Grassed waterways are used as
(a) Diversion channels (b) Outlets
(c) Inlet to terrace system (d) All of these

50. Broad base terrace is also known as
(a) Bench terrace (b) Channel terrace
(c) Drainage type terrace (d) None of these

51. Scattering of detached soil particles by the impact of rain drops is called
(a) Gully erosion (b) Ravines
(c) Sheet erosion (d) Splash erosion

52. Muddy runoff from the field is an indication of
(a) Splash erosion (b) Sheet erosion
(c) Rill erosion (d) Ravines

53. Which among the followings is erosion permitting crop?
(a) Maize (b) Sorghum
(c) Cotton (d) All the above

54. Which among the followings is erosion resisting crop?
(a) Grasses (b) Legumes
(c) Cereals (d) Both a and b

55. Which is the most popular soil conservation measure adopted in India?
(a) Bench terracing (b) Contour bunding
(c) Contour strip (d) All of these

56. Which of the following types of erosion can not be prevented by normal tillage operations?
(a) Rill erosion (b) Gully erosion
(c) Sheet erosion (d) Splash erosion

57. Based on time of occurrence drought has been classified into ______ kinds.
(a) 2 (b) 3 (c) 4 (d) 5

58. Desert climate is the characteristics of __________ drought

(a) Invisible (b) Climatic
(c) Seasonal (d) Permanent

59. Which one of the following chemicals is used as an antitranspirant?

(a) Diuron (b) Paraquat
(c) Glyphosate (d) Cycocel

60. Which one of followings is most popular dryland crop?

(a) Paddy (b) Maize
(c) Sorghum (d) None of these

61. Cycocel is a ________________ type of antitranspirant

(a) Stomatal closure (b) Growth retardant
(c) Film forming (d) None of these

62. Cooling of leaf surface with white material increases

(a) Albedo (b) Transpiration
(c) Growth (d) Photosynthesis

63. Dust mulch is effective in

(a) Coarse textured soil (b) Heavy textured soil
(c) Both (a) & (b) (d) None of these

64. Contour trenches are recommended for slopes between

(a) 5.0-6.0 % (b) 7.0-8.0 %
(c) 10.0-16.0% (d) 20-30 %

65. Which among the followings is channel stabilization measure?

(a) Graded bunding (b) Bench terracing
(c) Gabion check dams (d) V-ditch

66. Surface runoff water harvesting is achieved through

(a) Contour bunding (b) Levelling
(c) Deep ploughing (d) Dugout ponds

67. Stream flow runoff harvesting is achieved through

(a) Check dams (b) Dugout ponds
(c) Khadins (d) All of these

68. Diameter of soil particles most susceptible to wind erosion is

(a) 0.1 mm (b) 0.2 mm (c) 0.02 mm (d) 0.1 mm

69. Tankas and Khadins are water harvesting structures generally constructed in

(a) North – West Himalayas
(b) North-East Himalayas
(c) Arid and Semi-arid areas of Rajasthan
(d) Western and Central India

70. Bottom Width of gabion walls/ retaining walls for slope stabilization is roughly kept as

(a) One fourth of height of wall (b) One third of height of wall
(c) One half of height of wall (d) Two third of height of wall

71. For vegetative barriers in semi arid regions of central India which grass species is most suitable?

(a) Guinea (b) Napier (c) Khus (d) Dhaman

72. Out of total number of land capability classes, number of land classes suitable for cultivation is

(a) Eight (b) Four (c) Six (d) Two

73. Neeranchal watershed programme was implemented in the year

(a) 2010 (b) 2012 (c) 2015 (d) 2018

74. Wind and water erosion contribute what percentage of soil degradation ?

(a) 50% (b) 60% (c) 70% (d) 80%

75. The maximum impact of raindrop will be at what angle when it hits the ground?

(a) 30° (b) 45° (c) 90° (d) 180°

76. Which of the following soil separates has more erodability ?

(a) Sand (b) Silt (c) Clay (d) Rocks

77. Drought Prone Area Programme was started in

(a) 1965 (b) 1970 (c) 1973 (d) 1999

78. Desert Development Programme was initiated in

(a) 1970 (b) 1977 (c) 1980 (d) 1985

79. Steeply inward slopping narrow bench terraces constructed on contour is called

(a) Gradonies (b) Graded bunding
(c) V-ditch (d) Trenching

80. Contour bunding is recommended for slopes less than

(a) 2% (b) 4% (c) 6% (d) 16%

Answer Keys

Q.No.	Ans.	Q.No.	Ans.	Q.No.	Ans.	Q.No.	Ans.	Q.No.	Ans.	Q.No.	Ans.
1	(c)	15	(a)	29	(a)	43	(a)	57	(b)	71	(a)
2	(a)	16	(a)	30	(b)	44	(a)	58	(d)	72	(b)
3	(b)	17	(b)	31	(a)	45	(a)	59	(d)	73	(c)
4	(c)	18	(c)	32	(a)	46	(a)	60	(c)	74	(a)
5	(c)	19	(a)	33	(b)	47	(c)	61	(b)	75	(c)
6	(b)	20	(a)	34	(c)	48	(d)	62	(a)	76	(b)
7	(a)	21	(a)	35	(a)	49	(b)	63	(b)	77	(c)
8	(a)	22	(b)	36	(d)	50	(c)	64	(d)	78	(b)
9	(a)	23	(a)	37	(b)	51	(d)	65	(c)	79	(a)
10	(c)	24	(b)	38	(a)	52	(b)	66	(d)	80	(c)
11	(a)	25	(c)	39	(c)	53	(d)	67	(a)		
12	(b)	26	(a)	40	(a)	54	(d)	68	(a)		
13	(c)	27	(b)	41	(c)	55	(b)	69	(c)		
14	(a)	28	(b)	42	(b)	56	(b)	70	(d)		

II. Fill in the blanks

1. All India Coordinated Research Project on Dryland Agriculture (AICRPDA) was launched in ___________. **[1970]**
2. Central Research Institute for Dryland Agriculture (CRIDA) is located at ___________. **[Hyderabad]**
3. CRIDA was established in ___________. **[1985]**
4. Moisture Deficit Index (MDl), for semi-arid climate is ___________. **[(–) 33.3 to (–) 66.7]**
5. ICRISAT was established in ___________. **[1972]**
6. NWDPRA (National Watershed Development Programme for Rainfed Agriculture) was launched in ___________. **[1985-86]**
7. Out of total cultivated land in India about ___________% is under dry land agriculture. **[65]**
8. The coefficient of variation ___________ with the decease in annual rainfall. **[Increase]**
9. Based on average annual rainfall, high rainfall zones are those where rainfall is ___________. **[>1150 mm]**
10. ___________ is the most limiting factor in dryland agriculture. **[Soil moisutre]**
11. As moisture stress increases turgidity of guard cells ________. **[Decreases]**
12. A farm pond of ________ m^3 capacity is sufficient for one hectare of catchment area in black soils. **[150]**
13. Rainfed agriculture in India contributes _______ % food and supports _______% of the population. **[44, 40]**
14. Dry land farming areas are grouped under ___________ Regions. **[Arid and semi-arid]**
15. Rainfed farming is practiced in ___________ regions. **[Humid]**
16. In tropical climate, even the coolest month has an average temperature more than _____°C. **[18]**
17. If the average temperature of warmest month is less than 10°C it is called ___________ climate. **[Polar]**
18. If the annual rainfall is less than evaporation, it is called ___________ climate. **[Dry Climate]**
19. Thornthwaite classified climate based on ___________ Index. **[PE]**
20. Based on MDI, the value of MDI for sub-humid climate is _________. **[0–33]**
21. Watershed is synonym of ___________. **[Drainage basin or Catchonent area]**

22. Central Soil and Water Conservation Research and Training Institute, Dehradun was established in __________. **[1974]**

23. CAZRI is situated at __________. **[Jodhpur]**

24. CAZRI was established in __________. **[1959]**

25. Desert Development Programme was started in __________. **[1977]**

26. 20 million ha land lies in __________ rainfall belt. **[>1150 mm]**

27. 15 million ha land lies in __________ mm rainfall zone. **[500–750]**

28. Of the total dry land area __________% is covered by vertisols and associated soils. **[35]**

29. Contour bunding is recommended for rolling lands with slope of __________%. **[less than 6]**

30. Graded bunds is highly suitable for area with annual rainfall of more than __________ mm. **[800]**

31. For seed hardening of paddy __________ is used. **[KCl]**

32. In sorghum __________ is used for seed hardening. **[Potassium dihydrogen phosphate (KH_2PO_4)]**

33. Period of seed hardening varies from __________ hours depending on the nature of seed coat. **[6–18]**

34. Kaolin, as an antitranspirant, is used at ______ % concentration. **[3–6]**

35. Cycocel is a __________ type of antitranspirant. **[Growth retardant]**

36. The size of model watershed varies from __________ ha. **[200–2000]**

37. Most widely accepted climatic classification was given by __________. **[Koppen and Thornthwaite]**

38. __________ accumulation in plants is considered as an indicator of water stress. **[Proline]**

39. __________ is used as growth retardant for the purpose of increasing yield under dry lands. **[Cycocel]**

40. Water stress results in __________ of osmotic potential in plants. **[Reduction]**

41. __________ is most effective means of reducing transpiration in field. **[Weed control]**

42. Plants which germinate at the beginning of rainy season and have an extremely short life span of 4-6 weeks is called __________. **[Ephemerals]**

43. International Institute of Tropical Agriculture is situated in __________. **[badan, Nigeria]**

44. Aridity index for sub-humid regions lies between __________. **[0.5 to 0.65]**

45. For probability of rainfall analysis atleast ____ years data are required. **[30]**

46. Coefficient of rainfall variation must be within ____% for its reliability. **[20]**

47. Aridity index for evaluating drought by using water balance was proposed by __________. **[Thronthwaite]**

48. ________________ is used as subsurface barrier in dryland agriculture. **[Bentonite clay]**

49. Erosion permitting crops alternating with erosion resisting crop is called __________ crops. **[contour stripping]**

50. Graded bunds are also called as __________. **[channel terracing]**

51. Planting tall growing crops alternate with low growing crops is known as __________ cropping. **[Winds strip]**

52. The most efficient and chepeast method of conserving rainfall is to hold water __________. **[*in-situ*]**

53. Runoff in tropical and subtropical climate varies from __________ % of total rainfall. **[10–40]**

54. Ratio of soil, cowdung and straw as sealants in farm pond is kept ______________. **[10:1:1]**

55. Contingent management practice to overcome unexpected weather condition is known as ________________. **[Midterm correction]**

56. Rotation of legume forage with cereals is known as ______________.

[Ley farming]

57. *Stylosanthes hamata* is a good example of __________. **[Legume pasture]**

58. The crops suitable for arid area of northern India during *kharif* are ________ ________________________________.

[Pearlmillet, Greengram, Mothbean and Sesamum]

59. Chronically drought affected area of India is ______________________ ________________________________.

[Desert Rajasthan and Kutch area of Gujarat]

60. Period of recurrence of drought is minimum in ___________. **[Assam]**

III. True or False

1. Early seeding is extremely important in dryland areas. **[T]**
2. A principle feature of dry lands is a negative water balance. **[T]**
3. There is a negative relationship between occurrence and strength of crusts with the content of clay and silts. **[F, Positive]**
4. By and large soils in dryland regions in India are deficient in N and P. **[T]**
5. Drylands are more thirsty than hungry. **[T]**
6. Antitranspirants are used to control evaporation. **[F, transpiration]**
7. Any material used at the soil surface for covering with a primary objective of reducing the evaporation loss is known as anti-evaporant. **[F, Mulch]**
8. Any material applied to plant surface to reduce water loss is known as mulch. **[F, anti-transpirants]**
9. Anti transpirant increases plant water potential. **[T]**
10. Reflecting type of anti-transpirants increases leaf temperature. **[F, Decreases]**
11. The film forming and stomata closing type of anti-transpirants increases leaf temperature. **[T]**
12. Application of sodium salts to soil increases runoff from treated catchment. **[T]**
13. Natural clay soil linings are generally the most eco-nomical in controlling the seepage losses. **[T]**
14. Square or rectangular ponds are more feasible for harvesting the runoff water than circular ponds. **[T]**
15. The ideal location of pond is in the corner of cultivated field. **[F, middle]**
16. In dryland agriculture crop failures are rare and drain-age is the important problem. **[F, Rainfed farming]**
17. In rainfed farming water erosion is important constraint. **[T]**
18. Regions with average temperature of coolest month in the range of 18°C to –3°C are grouped under temperate climate. **[T]**
19. Koppen classified climate into 10 groups. **[F, 5]**
20. CAM plants are most drought resistant plants. **[T]**
21. To improve infiltration of rain water in black soils verical mulches are useful. **[T]**
22. Phenyl Mercuric Acetate (PMA) is a stomatal closing type of anti-transpirant. **[T]**
23. Hexadecanol and silicon are film forming type of anti-transpirant. **[T]**
24. Soils of dry regions are deficient in nitrogen. **[T]**

25. Wind breaks obstruct wind flow and reduces wind speed. [T]
26. Shelterbelts are rows of trees planted for protection of crops against wind. [T]
27. When soil moisture and rainfall are inadequate during growing season to support crop growth and cause crop stress and wilting, it is called agricultural drought. [T]
28. Deficiency of water for hydrological purposes or for an industry which uses a large volume of water is known as hydrological drought. [T]
29. Deficiency of water during the crop growing season to meet the evapotranspiration demand is known as meteo-rological drought. [T]
30. Plants shows deficiency symptoms of water even when there is sufficient water present in soils is termed as physiological drought. [T]
31. In soils of very shallow depth (<7.5 cm) contour bunding is not suitable. [T]
32. Medium deep black soils are quite suitable for bunding. [T]
33. In deep black soils contour bunds are recommended. **[F, Not recommended]**
34. Graded bunds are ideally suitable for most of the red soils. [T]
35. Graded bund is restricted to 6% slope. [T]
36. Bench terracing is suitable for hilly areas with a slope of 6 to 33%. [T]
37. Levelled bench terrace is suitable for areas having medium rainfall with uniform distribution. [T]
38. Inward sloping bench terraces are most effective for heavy rainfall areas. [T]
39. Compartmental bunding is the cheapest and easily manipulable soil moisture conservation technique for rain fed Vertisols of less than 1% **slope.** [T]
40. Repeated soaking and drying of seeds is important to induce drought tolerance. [T]
41. Depth of sowing seed in dryland agriculture is kept more compared to irrigated agriculture. [T]
42. In dryland farming seed rate is kept normal as that in irrigated crops. **[F, 25% more]**
43. Thinning of plant stand is recommended to reduce the severity of moisture stress. [T]
44. Soil mulch effectively controls evaporation by breaking capillary continuity. [T]
45. Atrazine in low concentration can be used as anti-transpirant. [T]
46. Atrazine is a film forming type of antitranspirant. **[F, Stomatal closing type]**
47. Use of antitranspirant reduces photosynthesis. [T]

48. Dug-out farm ponds are constructed for surface runoff water harvesting. [T]

49. Under All India Coordinated Research Proiect on Dryland Agriculture moisture deficit index is taken into consi-deration for climatic classification. [T]

50. Red soils occupies the highest area under dryland agri-culture. **[F, Black soils]**

51. Awned crop varieties are more drought tolerant. [T]

52. Agricultural drought frequently referred as soil drought. [T]

53. Intercropping is more assured system than double cropping in drylands of India. [T]

54. Circular farm ponds has higher water storing capacity than other shapes. [T]

55. Dug-out farm ponds are a common feature on hill slopes. **[F, Flat terrains]**

56. Dhaman, anjan and sewan grasses are suitable for vegetative barriers in arid region. [T]

57. As per irrigation commission, Govt. of India and Indian meteorological Department drought years have been defined as those years for which annual rainfall is less than or equal to 75% of normal. [T]

58. Farmers in drylands rarely use fertilizers. [T]

59. The ratio of annual precipitation to potential evapotrans-piration is less than 0.05 for hyper arid. [T]

60. According to growing period concept of FAO, areas having a growing period between 1 and 74 days are classified as arid. [T]

61. Growing period between 75 and 119 days are cate-gorised as semiarid. [T]

62. Famine commission was appointed in 1980. [T]

63. Under dry land farming, fertilizers are applied at higher rates. **[F, Moderate]**

64. Tropical dry lands have lesser carry over of moisture from one season to other than temperate dry lands. [T]

65. The intensity of rainfall in tropical dry land is very low than temperate dry lands. **[F, Very high]**

66. The value of coefficient of variation of rainfall is higher at Jodhpur than Hyderabad. [T]

67. Vertisols are called self ploughing soils. [T]

68. Interrow water harvesting system is a practice suitable for light textured soils. [T]

69. Use of moderate dose of fertilizer increases moisture use efficiency in drylands. [T]

70. Scoops are used for soil and water conservation *in situ*. [T]

IV. Match the followings

1. Match the years with significance in dryland agriculture

1.	1953	a	Establishment of Operational Research Project on Dryland Agriculture
2.	1976	b	Established Central Soil Conservation Board
3.	1983	c	Started 47 model watersheds under ICAR
4.	1985	d	Inaguration of CRIDA at Hyderabad

2. Match the terms with suitable explanation

1.	Soil drought	a	Depletion of surface water and consequent drying of reservoirs
2.	Hydrological drought	b	Abnormal failure of rainfall specially in humid or sub humid climate, brief and irregular
3.	Contingent drought	c	Imbalance between available soil moisture and evapotranspiration of crop
4.	Atmospheric drought	d	Low air humidity accompanied by hot dry winds

3. Match the soil moisture conservation measures with its suitability

1.	Dead furrows at 3.6 m interval	a	Red soil with high rainfall
2.	Sowing in flat and ridging later	b	Red soil with low rainfall or black soil with medium rainfall
3.	Graded border strips	c	Black soil with low rainfall
4.	Compartmental bunding	d	Red soil with medium rainfall

4. Match different types of mulches with its suitable explanation

1.	Soil mulch	a	Vertical trenches filled with crop residues reduce evaporation
2.	Vertical mulch	b	Polyethylene or polyvinylchloride reduce evaporation
3.	Plastic mulch	c	Paddy or wheat straw reduces evaporation
4.	Straw mulch	d	Intercultivation creates soil mulch and minimize evaporation

5. Match the dryland implements with their use

1.	CRIDA drill plough	a	Seed cum fertilizer drill for ragi consists of five furrow openers
2.	CRIDA seed cum seeder plough	b	Two row bullock drawn fertilizer drill

3. Ridger seeder	c	Seed cum fertilizer placement device attached with country
4. Ragi seeder	d	Three row seed cum fertilizer drill

6. Match the land capability class with its suitable farming system

1. II and III	a	Moderately intensive cultivated alley cropping, agro-horticulture, agri-silviculture
2. IV	b	Recreation, wild life, watershed protection
3. V and VI	c	Horti/silvi-pastoral *system*
4. VIII	d	Restricted cultivation-Ley farming

7. Match mean annual temperature with its temperature belt

1. 28° C or more	a	Very cold
2. 25 to 28°C	b	Mild
3. 20 to 25°C	c	Hot
4. 10°C or less	d	Very hot

8. Match moisture index value with its moisture belt

1. < 80	a	Slightly moist
2. (-) 40 to (–) 60	b	Wet
3. 0 to (–) 20	c	Dry
4. (+) 50 to (+) 100	d	Extermely dry

9. Match the authors with the criteria for climatic classification

1. Thornthwaite and Mather	a	Mean annual precipitation and mean annual PET
2. Krishnan and Mukhtar Singh	b	Monthly rainfall distribution
3. National Commission on Agriculture	c	Monthly precipitation, PET and soil moisture storage
4. Troll	d	Duration of humid months based on monthly rainfall

10. Match Koppen's classification of climate with their characters

1. Tropical rainy climate	a	Tropical deserts have average annual temperature exceeding 18°C
2. Dry climate	b	Temerature of coldest month exceeding 18°C
3. Warm temperate rainy climate	c	Average temperature of warmest month less than 10°C
4. Polar climate	d	Average temperature of coldest month less than 18°C but greater than (-) 30 °C

11. Match biophysical induced stress with suitable example

1.	Chemical conditions	a	Accelerated soil erosion, soil compaction
2.	Physical conditions	b	Floods and droughts, earth quake
3.	Catastrophic events	c	Oxygen deficiency, acidification
4.	Holistic conditions	d	Poor soil health

12. Match rainfall zones and soil types with land treatment

1.	Rainfall < 500 mm and aridisols	a	Raised beds and suncken system
2.	Rainfall 500-1000 mm and red	b	Interplot water harvesting of soils 1:1 cropped to uncroped land
3.	Rainfall > 1000 mm and alluvial soils	c	Dead furrows at 3-6 m interval
4.	Rainfall > 1000 mm and black	d	Sowing and ridging soils

13. Match dry-farming ecozones with their characters

1.	Arid sub-tropics	a	Annual rainfall 900 mm, ET 1800, LGP 120-180 days, double cropping
2.	Semi-arid sub-tropics	b	Annual rainfall 350 mm, annual ET 2000 mm, LGP 70-90 days crops are grown on conserved soil moisture
3.	Sub humid tropics	c	Rainfall 500 – 750 mm, LGP 90-180 days, intercropping or monocropping
4.	Semi-arid tropics	d	Rainfall 1200 mm, LGP 210-300 days, intercropping or double cropping

14. Match the implement from which they were developed

1.	Ridger seeder	a	AICRPDA, Hisar
2.	Rayala Gorru	b	AICRPDA, Varanasi
3.	Malviya seed-cum-fertilizer drill	c	AICRPDA, Sholapur
4.	Shivaji multi-purpose farming machine	d	AICRPDA, Anantpur

15. Match the place with prevalent soil type

1.	Hisar	a	Alfisol
2.	Hyderabad	b	Aridisol
3.	Dehradun	c	Vertisol
4.	Indore	d	Inceptisol

16. Match the crops with commonly used seed hardening chemicals

1.	Cotton	a	Potassium dihydrogen
2.	Sorghum	b	$MgSO_4$
3.	Groundnut	c	$CaCl_2$
4.	Greengram	d	Succinic acid

17. Match the crops with per cent area under dryland conditions in India

1.	Maize	a	95%
2.	Pearlmillet	b	100%
3.	Pulses and oilseeds	c	80%
4.	Jute and mesta	d	75%

18. Match the type of land degradation with its extent in India

1.	Water erosion	a	10.46%
2.	Wind erosion	b	3.20%
3.	Waterlogged	c	2.38%
4.	Shifting cultivation	d	57.16%

19. Match the type of water erosion with its explanation

1.	Sheet erosion	a	Second stage of erosion when sheet erosion is allowed minute finger shaped grooves over field
2.	Rill erosion	b	Initial stage of soil erosion, removal of thin venier of soil
3.	Ravines	c	Deep and wide gullies formed, advanced stage of gully erosion
4.	Land slides	d	Sliding down of large chunk of soil due to steep slopes

Answer Keys

Q.No.	Ans.	Q.No.	Ans.
1	b, a, c, d	11	c, a, b, d
2	c, a, b, d	12	b, d, c, a
3	b, d, a, c	13	b, a, d, c
4	d, a, b, c	14	a, d, b, c
5	c, b, b, a	15	b, a, d, c
6	a, d, c, b	16	d, a, c, b
7	d, c, b, a	17	c, a, d, b
8	d, c, a, b	18	d, a, b, c
9	c, a, b, d	19	b, a, c, d
10	b, a, d, c		

◆

Appendix – I

Acronyms

AICRPDA	All India Coordinated Research Project on Dryland Agriculture
BBF	Broad Bed and Furrow
CAZRI	Central Arid Zone Research Institute
CIDA	Canadian International Development Agency
CRIDA	Central Research Institute for Dryland Agriculture
DDP	Desert Development Programme
DPAP	Drought Prone Area Programme
ET	Evapo-transpiration
IAEDS	Intergrated Afforestation and Eco- Development Scheme
ICAR	Indian Council of Agricultural Research
ICRISAT	International Crop Research Institute for Semi Arid Tropics
IMD	India Meteorological Department
IWDP	Intergrated Watershed Development Programme
IWMP	Intergrated Watershed Management Programme
LTA	Long Term Average
NRAA	National Rainfed Area Authority
NWDPRA	National Watershed Development Programme for Rainfed Areas
ORP	Operational Research Project
PET	Potential Evapo-transpiration
PIA	Project Level Implementing Agency
PMA	Phenyl Mercuric Acetate
PRT	Puerto Rico Terrace
SLNA	State Level Nodal Agency
SWT	Stone Wall Terracing
WC	Watershed Committee
WCDC	Watershed Cell Cum Data Centre
WDC- PMKSY	Watershed Development Component of Prime Minister Krishi Sinchayee Yojna
WDP	Watershed Development Project

Appendix - II

International/ National Research Institutes related to Dryland Agriculture

1	ICAR – Central Institute of Arid Horticulture (CIAH)	Bikaner, Rajasthan
2	ICAR – Central Arid Zone Research Institute (CAZRI)	Jodhpur, Rajasthan
3	ICAR- Central Research Institute for Dryland Agriculture (CRIDA)	Hyderabad, AP
4	ICAR- Indian Institute of Soil and Water Conservation (IISWC)	Dehradun, Uttarakhand
5	ICAR- National Institute of Abiotic Stress Management (NIASM)	Malegaon, (Maharashtra)
6	ICAR – Central Agro-foresty Research Institute (CAFRI)	Jhansi, UP
7	International Center for Agricultural Research in the Dry Areas (ICARDA)	Alleppo, Syria
8	International Centre for Research in Agro Forestry (ICRAF)	Nairobi, Kenya
9	International Crops Research Institute for Semi Arid Tropics (ICRISAT)	Hyderabad, India
10	International Institute of Tropical Agricultural (IITA)	Lbadan, Nigeria

Appendix - III

References Cited

Dhopte, Arvind M. 2002. Agro Technology for Dryland Agriculture. Scientific Publishers, New Pali Road P.O. Box. 91, Jodhpur - 342001

Dhurva Narayanm, V.V., Sastry, G. and Patanik, U.S. 1997. Watershed Management. Krishi Anusandhan Bhawan, ICAR, New Delhi - 110012

Govindan., K. and Thirumurugan. 2003. Principles and Practices of Dryland Agriculture. Kalyani Publishers, New Delhi - 110002

Gupta, U.S. (Ed). 1975. Physiological Aspects of Dryland Farming. Oxford and IBH Publishing Co. New Delhi - 313002

Jat, M.L., Singh. P., Bhakar. S.R. and Sharma S.K. Drought Management. 2010. Kalyani Publishers New Delhi- 110002

Jat. M.L., Bhakar. S.R., Sharma S.K. and Kothari. A.K. 2013. Scientific Publishers, New Pali Road P.O. Box. 91, Jodhpur - 342001

Kadale A.S. 2014. Integrated Watershed Approach for Natural Resources Management. Agrotech Publishing Academy, Hiran Magri, Udaipur - 313002

Khan. M.A. 2012. Watershed Management for Sustainable Agriculture. Agrobios (INDIA), Jodhpur- 342003

Mahnot. S.C., Singh. P.K. and Chaplot. P.C. 2012.Soil and Water Conservation & Watershed Management. Apex Publishing House, Udaipur - 313002.

Maliwal. G.L. and L.L. Somani. 2015. Text Book of Dryland Farming. Agrotech Publishing Academy, Hiran Magri, Udaipur- 313002

Oswal. M.C. 2016. Watershed Management for Dryland Agriculture. Associated Publishing Company, New Delhi - 110002.

Panda S.C. 2018. Dryland Agricultural. Agrobios (India) Jodhpur - 342003

Reddy, S.R., Prabhakara Reddy, G. 2018. Rainfed Agriculture and Watershed Management. Kalyani Publishers, New Delhi - 110002

Sharma, Rohit and Sharma Vikas. 2018. Rainfed Agriculture. Bioscientific Publishers. New Delhi.

Singh Pratap and Maliwal. P. L. 2005. Technologies for Food Security and Sustainable Agriculture. Agrotech Publishing Academy, Hiran Magri, Udaipur - 313002.

Singh Surjit and Rathore M.S. 2010. Rainfed Agriculture in India. Rawat Publications, Jaipur.

Singh, Rajvir. 2003. Watershed Planning and Management. Yesh Publishing House, Bikaner – 334003

Singh. R. P. 2016. Sustainable Development of Dryland Agriculture in India Scientific Publishers, New Pali Road P.O. Box. 91, Jodhpur, India - 342001

Venkeateswarlu, J. 2003. Rainfed Agriculture in India – Research and Development Scenario Indian Council of Agriculture Research, New Delhi - 110012

■